Springer Tracts in Mechanical Engineering

Springer Tracts in Mechanical Engineering (STME) publishes the latest developments in Mechanical Engineering - quickly, informally and with high quality. The intent is to cover all the main branches of mechanical engineering, both theoretical and applied, including:

- Engineering Design
- Machinery and Machine Elements
- Mechanical Structures and Stress Analysis
- Automotive Engineering
- Engine Technology
- Aerospace Technology and Astronautics
- Nanotechnology and Microengineering
- Control, Robotics, Mechatronics
- MEMS
- Theoretical and Applied Mechanics
- Dynamical Systems, Control
- Fluids Mechanics
- Engineering Thermodynamics, Heat and Mass Transfer
- Manufacturing
- Precision Engineering, Instrumentation, Measurement
- Materials Engineering
- Tribology and Surface Technology

Within the scope of the series are monographs, professional books or graduate textbooks, edited volumes as well as outstanding PhD theses and books purposely devoted to support education in mechanical engineering at graduate and post-graduate levels.

Indexed by SCOPUS, zbMATH, SCImago.

Please check our Lecture Notes in Mechanical Engineering at http://www.springer.com/series/11236 if you are interested in conference proceedings.

To submit a proposal or for further inquiries, please contact the Springer Editor **in your country**:

Dr. Mengchu Huang (China)
Email: mengchu.Huang@springer.com
Priya Vyas (India)
Email: priya.vyas@springer.com
Dr. Leontina Di Cecco (All other countries)
Email: leontina.dicecco@springer.com

All books published in the series are submitted for consideration in Web of Science.

More information about this series at http://www.springer.com/series/11693

Shun-Qi Zhang

Nonlinear Analysis of Thin-Walled Smart Structures

Springer

Shun-Qi Zhang
School of Mechatronic
Engineering and Automation
Shanghai University
Shanghai, China

ISSN 2195-9862 ISSN 2195-9870 (electronic)
Springer Tracts in Mechanical Engineering
ISBN 978-981-15-9856-2 ISBN 978-981-15-9857-9 (eBook)
https://doi.org/10.1007/978-981-15-9857-9

This Springer imprint is published by the registered company Springer Nature Singapore Pte Ltd.
The registered company address is: 152 Beach Road, #21-01/04 Gateway East, Singapore 189721, Singapore

Dedicated to my beloved wife, daughter, parents and sisters.

Acknowledgements

The work presented in the report was started since 2010 when I was taking the Ph.D. study at RWTH Aachen University (Germany). After finishing my Ph.D. degree, I worked as an Associate Professor in Northwestern Polytechnical University (P.R. China) and later in Shanghai University (P.R. China).

The work was under the financial support from China Scholarship Council during 2010–2014. Then it was supported by the National Natural Science Foundation of China (Grand Nos. 11972020, 11602193).

First, I wish to express my deepest gratitude to Prof. Dr.-Ing. Rüdiger Schmidt, the former Vice Director of the Institute of General Mechanics, RWTH Aachen University. He supervised my Ph.D. research during 2010–2014 and collaborated with my research work till today. I would like to express my sincere gratitude to Univ.-Prof. Dr.-Ing. Dieter Weichert, the former Head of the Institute of General Mechanics. He supported me greatly on the research during my Ph.D. study. I want to express my appreciation to Prof. Dr. Xiansheng Qin, the former Head of the Department of Industrial Engineering, Northwestern Polytechnical University, for his encouragement and support for my work not only during my Ph.D. study but also during my academic career in the university. I would also like to thank Prof. Yingjie Yu, the Dean of School of Mechatronic Engineering and Automation, Shanghai University, for her great support on my research in the field of smart structures.

Second, I would like to appreciate my students, who assisted my research in the field of smart structures, M.Sc. Faysal Andary, M.Sc. Haonan Li, M.Sc. Heyuan Wang, M.Sc. Yaxi Li, M.Sc. Shuyang Zhang, M.Sc. Yingshan Gao, M.Sc. Yafei Zhao, M.Sc. Ting Xue among many others.

Finally, I wish to thank my family for their unlimited support and encouragement on my work. I would also wish to thank all my friends who helped and encouraged me in China and abroad.

Shanghai, China
Shun-Qi Zhang

Contents

Acronyms

Operators

$\dot{\mathrm{W}}$	first-order time derivative, or velocity
$\ddot{\mathrm{W}}$	second-order time derivative, or acceleration
$\mathrm{W}_{,\alpha}$	spatial derivative with respect to Θ^{α}
$\mathrm{W}_{\vert\alpha}$	covariant derivative with respect to Θ^{α}
$\delta\mathrm{W}$	variational operator
$\Delta\mathrm{W}$	incremental operator
exp	exponential operator
rank	rank
W^{T}	transposition
W^{-1}	inverse
$\Vert\ \Vert$	Euclidean norm
$\vert\ \vert$	absolute value
$\cdot$	scalar product, or dot product
$\times$	vector product
$\otimes$	tensor product
[ ]	matrix
{}	vector
$\bar{\mathrm{W}}$	quantities in the deformed configuration
$\breve{\mathrm{W}}$	quantities in the material coordinate system
$\widehat{\mathrm{W}}$	normalized quantities

Symbols

X^i	Cartesian coordinate system
Θ^i	curvilinear coordinate system

R	position vector for an arbitrary point in the shell space
r	position vector for an arbitrary point at the mid-surface
g_i	covariant base vectors in the shell space
g^i	contravariant base vectors in the shell space
g_{ij}	covariant metric tensor in the shell space
g^{ij}	contravariant metric tensor in the shell space
a_i	covariant base vectors at the mid-surface
a^i	contravariant base vectors at the mid-surface
a_{ij}	covariant metric tensor at the mid-surface
a^{ij}	contravariant metric tensor at the mid-surface
δ_i^j	Kronecker delta
$\Gamma^{\lambda}_{\alpha\beta}$	Christoffel symbols of the second kind
$b_{\alpha\beta}$	covariant components of the curvature tensor
b^{α}_{β}	mixed components of the curvature tensor
e^{ijk}	permutation symbol
$\boldsymbol{u}$	displacement vector for an arbitrary point in the shell space
$\overset{0}{\boldsymbol{u}}$	translational displacement vector at the mid-surface
$\overset{1}{\boldsymbol{u}}$	rotational vector of the Θ^3-line
v_i	translational displacements in the shell space
$\overset{0}{v}_i$	translational displacements at the mid-surface
$\overset{1}{v}_i$	rotational displacements at the mid-surface
$\boldsymbol{\mu}$	shifter tensor
μ^i_j	components of the shift tensor
H	mean curvature of the surface
K	Gaussian curvature of the surface
V	volume of the parallelepiped spanned by the covariant base vectors
$\boldsymbol{F}$	deformation gradient tensor
$\boldsymbol{C}$	right Cauchy-Green tensor
$\boldsymbol{G}$	Riemannian metric tensor
ρ	mass density
Y	Young's modulus
υ	Poisson's ratio
G	shear modulus
$\boldsymbol{\varepsilon}$	Green-Lagrange strain vector
$\boldsymbol{\sigma}$	second Piola-Kirchhoff stress vector
$\boldsymbol{c}$	elasticity constant matrix
$\boldsymbol{e}, \boldsymbol{d}$	piezoelectric constant matrix
$\boldsymbol{\epsilon}$	dielectric constant matrix
$\boldsymbol{D}$	electric displacement vector
$\boldsymbol{E}$	electric field vector
T	kinetic energy
W_{int}	internal work

W_{ext} external work
${}^{m}\mathrm{C}$ configuration m, $m = 0, 1, 2$
$\boldsymbol{M}_{\mathrm{uu}}$ mass matrix
$\boldsymbol{C}_{\mathrm{uu}}$ damping matrix
$\boldsymbol{K}_{\mathrm{uu}}$ stiffness matrix
$\boldsymbol{K}_{\mathrm{u}\phi}$ piezoelectric coupled stiffness matrix
$\boldsymbol{K}_{\phi\mathrm{u}}$ piezoelectric coupled capacity matrix
$\boldsymbol{K}_{\phi\phi}$ piezoelectric capacity matrix
$\boldsymbol{K}_{\mathrm{ug}}$ geometrically induced stiffness due to mechanically induced stresses
$\boldsymbol{K}_{\phi\mathrm{g}}$ geometrically induced stiffness due to electrically induced stresses
$\boldsymbol{S}_{\mathrm{uu}}$ mechanically induced resultant stresses
$\boldsymbol{S}_{\mathrm{u}\phi}$ electrically induced resultant stresses
$\boldsymbol{F}_{\mathrm{ue}}$ total external force vector
$\boldsymbol{F}_{\mathrm{ub}}$ element body force vector
$\boldsymbol{F}_{\mathrm{us}}$ element surface force vector
$\boldsymbol{F}_{\mathrm{uc}}$ element concentrated force vector
$\boldsymbol{f}_{\mathrm{b}}$ body force vector of an arbitrary point in the shell space
$\boldsymbol{f}_{\mathrm{s}}$ surface force vector of an arbitrary point at the mid-surface
$\boldsymbol{f}_{\mathrm{c}}$ concentrated force vector of an arbitrary point at the mid-surface
$\boldsymbol{F}_{\mathrm{ui}}$ total in-balance force vector
$\boldsymbol{F}_{\mathrm{ut}}$ inertial in-balance force vector
$\boldsymbol{F}_{\mathrm{uu}}$ mechanically induced in-balance force vector
$\boldsymbol{F}_{\mathrm{u}\phi}$ electrically induced in-balance force vector
$\boldsymbol{G}_{\phi\mathrm{e}}$ total external charge vector
$\boldsymbol{G}_{\phi\mathrm{s}}$ surface charge vector
$\boldsymbol{G}_{\phi\mathrm{c}}$ concentrated charge vector
$\boldsymbol{G}_{\phi\mathrm{i}}$ total in-balance charge vector
$\boldsymbol{F}_{\phi\mathrm{u}}$ mechanically induced in-balance charge vector
$\boldsymbol{F}_{\phi\phi}$ electrically induced in-balance charge vector
$\boldsymbol{q}$ nodal displacement vector
ϕ_{a} actuation voltage vector applied on piezoelectric layer
ϕ_{s} sensor voltage vector output from piezoelectric layer

Abbreviations

AEH Asymptotic expansion homogenization
AFC Active fiber composite
ANS Assumed natural strain
CLT Classical lamination theory
CNT Carbon nanotube
DOF(s) Degree(s) of ereedom

EAS	Enhanced assumed strain
FE	Finite element
FI	Full integration
FOSD	First-order shear deformation
GPI	Generalized-proportional-integral
HOSD	Higher order shear deformation
LIN5	LINear shell theory with five parameters
LQG	Linear quadratic Gaussian
LQR	Linear quadratic regulator
LRT5	Fully geometrically nonlinear shell theory with five parameters
LRT56	Large rotation theory with six parameters expressed by five nodal DOFs
MEE	Magneto-electro-elastic structures
MFC	Macro-fiber composite
MRT5	Moderate rotation theory with five parameters
PI	Proportional-integral
PVDF	Polyvinylidene fluoride
PZT	Lead Zirconate Titanate
RVE	Representative volume element
RVK5	Refined von Kármán type nonlinear shell theory with five parameters
SOSD	Second-order shear deformation
SRI	Selectively reduced integration
TL	Total lagrangian
TOSD	Third-order shear deformation
URI	Uniformly reduced integration

List of Figures

List of Tables

Chapter 1
Introduction

Abstract The chapter first discusses the application background of smart structures and the definition of smart structures. Later, the history of smart structures, including various programs, is introduced. Finally, the objectives of the report and the outlines are addressed.

1.1 Background

Due to light-weight design, thin-walled structures made of isotropic materials or laminated by orthotropic materials are applied in many fields of technology, e.g. aeronautical [1, 2] and aerospace [3–5], civil and automotive engineering. Although thin-walled structures possess many beneficial properties, e.g. reduction of weight, less raw material, etc., they tend to be instable and sensitive to vibrations. To promote the structural performance with remaining light weight, thin-walled structures integrated with smart materials i.e. piezoelectrics, electrostrictives, magnetostrictives and shape memory alloys (SMA), are called *smart structures*. Smart structures have excellent performance on vibration control [2, 3], shape control [4, 5], noise and acoustic control [6, 7], energy harvesting [8–13] and health monitoring [1, 14, 15], among many others.

Smart structures in this report refer to those integrated with smart materials acting as sensors and actuators that can sense the changes of environment and measure the system states itself, based on which a control action can be implemented to make the structures perform in a desired way. The field of smart structures is a newly proposed concept and the studies are still on the way to the expected smart structures. Therefore, the definition of smart structures are not unique. In 1990s, a general framework of the definition of intelligent structures was proposed by Wada et al. [16]. They divided the development of smart structures into four levels. The first level systems include *sensory structures* and *adaptive structures*. The *sensory structures* "which

S.-Q. Zhang, *Nonlinear Analysis of Thin-Walled Smart Structures*, Springer Tracts in Mechanical Engineering, https://doi.org/10.1007/978-981-15-9857-9_1

possess sensors that enable the determination or monitoring of the system states or characteristics", and the *adaptive structures* "possess actuators that enable the alteration of system states or characteristics in a controlled manner". The sensory structures have sensors, but possess no actuators. Conversely, the adaptive structure possess actuators, but does not have sensors. Later, higher level structures, *controlled structures*, were defined as those with both sensors and actuators that are connected into a feedback architecture. In the third level structures, *active structures* were proposed and used frequently in the literature. It was defined as a subset of controlled structures with the sensors and actuators highly integrated into a host structure as one object [16]. In the literature, the structures named smart structures or intelligent structures are mostly referred to active structures. The highest level, *intelligent structures*, was defined as those where sensors and actuators are highly integrated into a feedback architecture which also includes control logic and electronics [16, 17]. Additionally, Rogers [18] defined *intelligent material systems* in biological engineering point of view as "those with intelligence and life features integrated in the micro-structure of the material system to reduce mass and energy and produce adaptive functionality".

1.2 History of Smart Structures

The concept of smart structures or intelligent structures was initiated in 1950s. In the 1970s, Claus [19] of Virginia Institute of Technology embedded optical fiber sensors into carbon fiber-reinforced composites, which made the structures have ability to sense stress and fracture damage. This is the first experiment of smart structures, called adaptive structures at that time.

From 1980s, various programs of smart structures were launched by United States government. For example, the United States carried out the research on active control of structural vibration of B-1 aircraft's panel, and then studied the vibration of F/A-18 aircraft's vertical tail [20, 21]. In 1984, the U.S. Army Scientific Research Bureau sponsored the research of rotorcraft technology. One year later, the U.S. government launched the research plan of smart structures, requiring the spacecraft to be adaptive. After the catastrophic fracture accident of Boeing 737 aircraft on April 28, 1988, the United States Congress realized that the aircraft should have a self diagnosis and timely prediction system to avoid similar accidents of aircraft in service. The congress forced the aircraft company to complete the smart aircraft concept design within three years.

During 1990s, various programs of smart structures were launched by United States, Europe, and Asia. In the United States, the fundamental research activities were carried out by the Department of Defense funding agencies, e.g. the Army Research Office (ARO), the Office of Naval Research (ONR), and the Air Force Office of Scientific Research (AFOSR), whereas the applications-oriented research activities were carried out by the Defense Advanced Research Project Agency (DARPA). Therefore, most of the early research programs in smart structures were first initiated by ARO and then supplemented by DARPA. The early major programs focused on smart structures are as follows.

URI program, initiated by ARO in 1992, is a multidisciplinary research program in smart structures. The program was headed respectively by the University of Maryland, the Virginia Polytechnic Institute and State University, and Rensselaer Polytechnic Institute.

SPICES (Synthesis and Processing of Intelligent Cost Effective Structures) was sponsored by the Advanced Research Project Agency (ARPA) from 1993 to 1995, and was led by McDonnell Douglas. Several different composite plates and trapezoidal rails containing a combination of piezoelectric actuators, fiber-optic sensors, SMAs, and piezoelectric shunts were tested for damping augmentation, frequency shifting, and active vibration control.

The program of ASSET (Applications for Smart Structures in Engineering and Technology) was set up to exploit the smart structures technologies within the European Union under the IMT (Industrial Materials and Technologies) research program. About fifty organizations from the United Kingdom, France, Germany, Italy, etc. participated with the principal objectives of providing a forum and funds for communication, infrastructure, and exchange of information among partners.

At the same period of 1990s, research institutions targeting on smart materials and structures were booming in United States, Europe, Japan, Korea, and China.

1.3 Objectives and Outline

Piezoelectric laminated smart structures are widely used for aerospace and automotive industries, as well as civil engineering. Due to the small thickness, thin-walled structures are sensitive to external excitations resulting in large deformations and large amplitude vibrations. Additionally, the low damping makes the structure with long period of vibration, which probably cause delamination or fatigue damage. Furthermore, to achieve large actuation forces for vibration suppression, smart structures are hopefully under strong electric field. Structures undergo large deformations may produce additional positive or negative stiffness. This nonlinear phenomena is defined as geometrically nonlinear. Analogously, structures under strong electric field may influence the structural stiffness positively or negatively, which here is defined as electroelastic materially nonlinear effect. To predict precisely the response of structures undergoings large displacements and under strong electric fields, these two nonlinear phenomena must be taken into account.

Concerning piezoelectric embedded plate and shell structures made of e.g. aluminum alloys, composite, functionally graded materials, under multi-physics coupled fields, the modeling technique is critical for structural design and it is a challenging stuff. This report mainly focuses on nonlinear analysis of piezoelectric laminated smart structures, which is organized into six major chapters.

In Chap. 2, an overview of the recent development of modeling techniques for piezoelectric embedded smart structures is presented. The investigation covers the introduction of through thickness displacement hypotheses in plates and shells; analysis of various geometrically nonlinear plate/shell theories; discussion of electroelastic material linear and nonlinear modeling; multi-physics coupled modeling techniques for piezo structures; modeling techniques of piezoelectric fiber composite bonded structures. and the vibration control of piezo smart structures.

In Chap. 3, we first introduce and compare the hypotheses that have been already developed, which is followed by the definitions of base vectors and geometric quantities in curvilinear coordinate system. Afterwards, the strain-displacement relations for large rotation theory with six parameters based on first-order shear deformation hypothesis are derived, as well as those for various geometrically nonlinear shell theories ranging from von Kármán type nonlinearity to full geometric nonlinearity.

Chapter 4 presents constitutive relations for multi-functional materials, including piezoceramics, piezopolymers, macro-fiber composites. First the fundamental theory of piezoelectricity in 3-dimensional space is presented for piezoelectric materials. To deal with fiber based piezoelectric materials, a coordinate transformation law is constructed between the structural coordinates and fiber coordinates. Afterwards, the constitutive relations of two typical MFC patches are developed with consideration of multi-layered structures. Finally, electroelastic coupled materially nonlinear constitutive equations are constructed for the simulation of piezoelectric materials under strong electric filed.

Chapter 5 develops electro-mechanically coupled nonlinear finite element (FE) models with large rotations for static and dynamic analysis of composite and piezoelectric laminated thin-walled structures. The large rotation theory has six independent kinematic parameters expressed by five nodal degrees of freedom (DOFs) using Euler angles to represent arbitrary rotations in structures. To demonstrate the effect of the proposed large rotation FE models, other simplified nonlinear FE models are developed as well. Those nonlinear models are linearized by Total-Lagrangian formulations. In the last part of this chapter, several numerical algorithms are introduced for solving the coupled static and dynamic equations.

In Chap. 6, the finite element simulations of isotropic piezoceramics or polymers integrated smart structures are presented. The chapter first deals with the validation test of the present large rotation FE models by several static benchmark problems, buckling and post-buckling analysis of alloys and composite laminated thin-walled structures. Later, the nonlinear FE models based on various geometrically nonlinear shell theories are applied to static and dynamic analysis of piezoelectric integrated smart structures. In the final part of the chapter, the simulations of electroelastic materially nonlinear analysis are investigated, in the case of smart structures under strong electric field.

In Chap. 7, the simulations of macro-fiber composite (MFC) laminated smart structures are presented. Two types of MFC patches including MFC-d31 and MFC-d33 are considered in the simulations. In order to verify the present FE model, validation tests are conducted through a cantilevered MFC plate. Later, linear analysis of MFC bonded structures with arbitrary piezo-fiber orientation angles are carried out and discussed. Furthermore, applying various geometrically nonlinear shell theories, multi-MFC bonded plates and shells are analyzed and compared with each other.

The last chapter, Chap. 8, summarizes the present work and outlines the scope of the future work.

References

1. X. Qing, A. Kumar, C. Zhang, I.F. Gonzalez, G. Guo, F.K. Chang, A hybrid piezoelectric/fiber optic diagnostic system for structural health monitoring. Smart Mater. Struct. **14**, S98–S103 (2005)
2. E.F. Sheta, R.W. Moses, L.J. Huttsell, Active smart material control system for buffet alleviation. J. Sound Vib. **292**, 854–868 (2006)
3. Z. Li, P.M. Bainum, Vibration control of flexible spacecraft integrating a momentum exchange controller and a distributed piezoelectric actuator. J. Sound Vib. **177**, 539–553 (1994)
4. H. Baier, Approaches and technologies for optimal control-structure-interaction in smart structures. Trans. Built Environ. **19**, 323–335 (1996)
5. B.N. Agrawal, M.A. Elshafei, G. Song, Adaptive antenna shape control using piezoelectric actuators. Acru Astronautica **40**, 821–826 (1997)
6. S.B. Choi, Active structural acoustic control of a smart plate featuring piezoelectric actuators. J. Sound Vib. **294**, 421–429 (2006)
7. M.C. Ray, R. Balaji, Active structural-acoustic control of laminated cylindrical panels using smart damping treatment. Int. J. Mech. Sci. **49**, 1001–1017 (2007)
8. J. Feenstra, J. Granstrom, H. Sodano, Energy harvesting through a backpack employing a mechanically amplified piezoelectric stack. Mech. Syst. Signal Process. **22**, 721–734 (2008)
9. R. Ly, M. Rguiti, S. D'Astorg, A. Hajjaji, C. Courtois, A. Leriche, Modeling and characterization of piezoelectric cantilever bending sensor for energy harvesting. Sens. Actuators A: Phys. **168**, 95–100 (2011)
10. X.R. Chen, T.Q. Yang, W. Wang, X. Yao, Vibration energy harvesting with a clamped piezoelectric circular diaphragm. Ceram. Int. **38S**, S271–S274 (2012)
11. A. Messineo, A. Alaimo, M. Denaro, D. Ticali, Piezoelectric bender transducers for energy harvesting applications. Energy Procedia **14**, 39–44 (2012)
12. K.B. Singh, V. Bedekar, S. Taheri, S. Priya, Piezoelectric vibration energy harvesting system with an adaptive frequency tuning mechanism for intelligent tires. Mechatronics **22**, 970–988 (2012)
13. L. Zhou, J. Sun, X.J. Zheng, S.F. Deng, J.H. Zhao, S.T. Peng, Y. Zhang, X.Y. Wang, H.B. Cheng, A model for the energy harvesting performance of shear mode piezoelectric cantilever. Sens. Actuators A: Phys. **179**, 185–192 (2012)
14. D. Mayer, H. Atzrodt, S. Herold, M. Thomaier, An approach for the model based monitoring of piezoelectric actuators. Comput. Struct. **86**, 314–321 (2008)
15. Z. Wu, X.P. Qing, F.K. Chang, Damage detection for composite laminate plates with a distributed hybrid PZT/FBG sensor network. J. Intell. Mater. Syst. Struct. **20**, 1069–1077 (2009)
16. B.K. Wada, J.L. Fanson, E.F. Crawley, Adaptive structures. J. Intell. Mater. Syst. Struct. **1**, 157–173 (1990)
17. P. Gaudenzi, *Smart Structures: Physical Behavior, Mathematical Modeling and Applications* (Wiley, 2009)
18. G.A. Rogers, Intelligent material system-the dawn of a new materials age. J. Intell. Mater. Syst. Struct. **4**, 4–12 (1993)
19. H.R. Clauser, Modern materials concepts make structure key to progress. Mater. Eng. **68**(6), 38–42 (1968)
20. R.W. Moses, Contributions to active buffetingvalleviation programs by the nasa langley research center, in *Structural Dynamics, & Materials Conference & Exhibit* (pages Paper No. AIAA–99–1318, 1999)
21. S.C. Galea, T.G. Ryall, D.A. Henderson, R.W. Moses, E.V. White, D.G. Zimcik, Next generation active buffet suppression system, in *AIAA/ICAS International Air and Space Symposium and Exposition: The Next 100 Y* (July 2003)

Chapter 2
Literature Review

Abstract This chapter gives an overview of modeling and simulation techniques for smart structures. First, the chapter starts with various through thickness hypotheses for beam, plate and shell structures. Later, the development history of geometrically nonlinear theories in composite thin-walled structures are discussed, which is followed by the implementation of those nonlinear shell theories in smart structures. For the case of smart structures under strong electric fields, electroelastic materially nonlinear modeling methods are presented. In order to give a deep understanding of the multi-physics coupled phenomenon, the modeling techniques for many recently developed types of smart structures are presented, including functionally graded smart structures, electro-thermo-mechanically coupled structures, magneto-electro-elastic composites, and macro-fiber composites. Finally, a literature survey on vibration control of piezoelectric structures is discussed for the applications of vibration and noise reduction.

2.1 Plate/Shell Hypotheses and Applications to Linear Analysis

Smart structures are usually formed as beam, plate and shell structures with integrated smart materials. Solid elements can be employed directly for finite element analysis of smart structures. Since there is no additional assumption on the geometry, a relatively accurate result can be obtained. However, solid elements have large number of degrees of freedom, resulting in high computational costs. Many researchers implemented solid elements into linear FE analysis of smart structures, e.g. Tzou and Tseng [1], Ha et al. [2], Dube et al. [3], Kapuria and Dube [4], Ray et al. [5], He [6], Sze et al. [7], Sze and Yao [8, 9], Kapuria and Kumari [10] among many others. Additionally, Yi et al. [11], Klinkel and Wagner [12, 13] developed geometrically nonlinear FE models using solid elements for static and dynamic analysis of piezoelectric smart structures.

S.-Q. Zhang, *Nonlinear Analysis of Thin-Walled Smart Structures*, Springer Tracts in Mechanical Engineering, https://doi.org/10.1007/978-981-15-9857-9_2

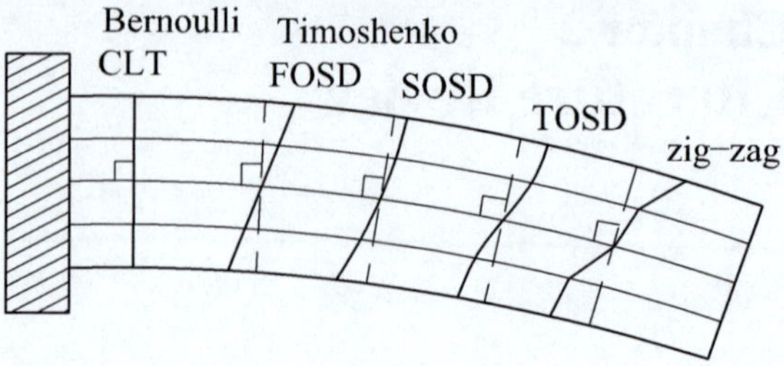

Fig. 2.1 Various hypotheses for plates and shells

Since the thickness is very small compared to the in-plane dimensions, thin-walled structures can be considered as 2-dimensional (2D) surfaces using proper through-thickness hypothesis, as shown in Fig. 2.1. The resulting elements are called plate or shell elements and the resulting method is 2D FE method. The through-thickness hypothesis is usually described as plate/shell theories in the literature, which defines the displacement distribution law through the thickness. Compared to solid elements, 2D plate and shell elements have the features of relatively high accuracy and less computational time. Such that plate and shell elements are frequently used in smart structures. In the plate/shell hypothesis, it assumes that the thickness remain constant during the structural deformation, by which the transverse normal strain is neglected. In addition, if one of the in-plane dimensions reduces to in the order of the thickness, the structures can be treated as a line using the Bernoulli or Timoshenko beam hypothesis. The resulting elements are called 1D line elements.

2.1.1 Kirchhoff-Love Hypothesis

The simplest plate/shell hypothesis is the Kirchhoff-Love hypothesis, known as classical plate/shell theory (CLT). The Kirchhoff-Love hypothesis assumes that a vector normal to the mid-surface in the undeformed configuration remains normal after deformation. A large number of papers were developed FE models with 2D element using the Kirchhoff-Love hypothesis. Tzou and Gadre [14], Lee [15] developed numerical models for PVDF bonded multi-layered thin plates and shells based on the Love's equation. Applying the Kirchhoff-Love hypothesis, Kioua and Mirza [16] constructed a linear finite element model for bending and twisting analysis of piezoelectric shallow shells. Many other studies have developed FE models for static and dynamic analysis of piezoelectric smart structures, see e.g. Lam et al. [17], Saravanos [18], Liu et al. [19]. Additionally, the classical plate theory was implemented into the analysis of vibration suppression using proportional feedback control [20] and optimal control [21].

From the assumption of classical plate/shell theory, the resulting numerical models neglect transverse shear strains. Due to the neglect of shear strains, a certain computational error may arise in the model. However, the error is negligible if the thickness are small enough. Therefore, the classical plate/shell theory is only valid for thin structures, rather than thick structures. Using the classical displacement distribution assumption, the resulting strain-displacement relations contain second-order derivative terms. Thus, it requires at least quadratic shape functions or higher-order elements in FE analysis.

By replacing the three parameters in the classical plate theory with two parameters, bending and shear components, yields a two-variable refined plate theory (RPT), which was developed by Shimpi and Patel [22]. The refined plate theory assumes that the vertical displacement consists of bending and shear components. The displacement distribution is a third-order function of the position in the thickness direction. The transverse shear strains are no longer zero or constant, but a second-order function of position in the thickness direction. Similar to the third-order shear deformation hypothesis, the transverse shear strains reach maximum at the mid-surface and disappear at outer surfaces.

2.1.2 *Reissner-Mindlin Hypothesis*

Due to neglect of the transverse shear strains, the classical plate/shell theory is only valid for thin structures. For moderately thick structures, transverse shear strains should be included in the model. Accounting for transverse shear strains, Reissner-Mindlin hypothesis was proposed and developed for plates and shells. The Reissner-Mindlin hypothesis, known as first-order shear deformation (FOSD) hypothesis, assumes that straight lines normal to the mid-surface remain straight after deformation, but not necessarily normal to the mid-surface. The FOSD hypothesis yields constant transverse shear strains through the thickness. For more details of the FOSD hypothesis for cylindrical and spherical shells, it refers to Ref. [23]. However, the consideration of constant transverse shear strains are not always valid for plates and shells, for example thick structures.

A large amount of publications have developed FE models based on the FOSD hypothesis for smart structure. The first analytical FOSD model of piezoelectric laminated plates was proposed and developed by Mindlin [24]. Later, the FOSD hypothesis was implemented into piezoelectric integrated smart structures for static analysis [25–28] and dynamic analysis [29–33]. Furthermore, a FOSD finite element model was developed by Wang [34] for piezoelectric bimorph structures. A meshfree model based on the FOSD hypothesis was developed by Liu et al. [35] for shape and vibration control of laminated composite plates.

2.1.3 *Higher-Order Shear Deformation Hypothesis*

The Kirchhoff-Love hypothesis is valid for thin structures, while the FOSD hypothesis is applicable for moderately thick structures. This is because the zero or constant transverse shear strains are not accurate enough for thick structures. The real situation of transverse shear strains is distributing nonlinearly through the thickness and disappearing at the outer surfaces. To model thick structures in a precise way, Reddy [36, 37] proposed a third-order shear deformation (TOSD), one of the higher-order shear deformation (HOSD) hypotheses, for composite laminated structures. The TOSD hypothesis assumes that the through-thickness displacement function is a third-order function of the position in the thickness direction. This yields second-order of the transverse shear strains, with the maximum shear strain at the mid-surface and zero shear strain at the outer surfaces. Afterwards, the theory was further applied to composite structures by Hanna and Leissa [38] and extended to model smart structures by Correia et al. [39, 40], Moita et al. [41], Selim et al. [42]. In addition, Loja et al. [43] and Soares et al. [44] proposed higher-order B-spline finite element models for composite structures laminated with piezoelectric patches.

2.1.4 Zigzag Hypothesis

Considering a laminated structure with different material properties, the inter-layer shear stresses are discontinuous when applying aforementioned plate or shell hypotheses. To avoid the inter-layer shear stress discontinuity, zigzag hypothesis or layerwise hypothesis was introduced. The hypothesis assumes that the displacement distribution function is different for each substrate layer, either with first-order or higher-order, in such a way the inter-layer shear stress continuity can be satisfied. A first-order zigzag shear deformation (or layerwise first-order shear deformation) theory was developed for smart structure by Ray and Reddy [45], Vasques and Rodrigues [46]. A third-order zigzag shear deformation theory was implemented into analysis of smart structures by Kapuria [47], Kapuria et al. [48]. Furthermore, Polit et al. [49] developed Murakami's zigzag formulation for modeling of laminated piezoelectric smart structures, while Carrera and Demasi [50] applied the theory for composite structures.

2.1.5 Bernoulli and Timoshenko Beam Hypotheses

Regarding to beam- or arch-shaped one-dimensional structures, they can be shrunk to a line for simplicity using specific beam hypothesis. Bernoulli and Timoshenko hypotheses are the most frequently used ones for mathematical modeling of beam-shaped structures. These two beam hypotheses were proposed earlier than plate and shell hypotheses. Therefore, the Kirchhoff-Love plate/shell hypothesis can be understood as an extension of the Bernoulli beam hypothesis. Analogously, the Reissner-Mindlin plate/shell hypothesis was extended from the Timoshenko beam hypothesis.

Neglecting the transverse shear strains, Crawley and Luis [51] proposed an analytical model for beam-like structures embedded with piezoelectric layer. Afterwards, Tzou and Chai [52], Kucuk et al. [53] developed linear models based on the Bernoulli beam hypothesis for vibration suppression of smart structures. Applying the Timoshenko beam hypothesis, Narayanan and Balamurugan [54], and Marinaki et al. [55] developed FE models for vibration control of smart structure, while Zu [56] investigated for energy harvesting.

2.2 Geometrically Nonlinear Modeling in Composites

2.2.1 Simplified Nonlinear Modeling

Geometrically linear models are only valid for structures undergoing small displacements and rotations. Geometrically nonlinear models were first developed for composite laminated structures or single layer monolithic structures. For simplicity, imposed with additional assumptions like small or moderate rotations, or weak nonlinear effect, yields various geometrically nonlinear theories, here called simplified nonlinear theories. The von Kármán type nonlinear theory is the simplest geometrically nonlinear theory, which only considers the nonlinear effect resulting from the transverse displacements and under the assumption of small rotations. A large number of publications can be found that developed von Kármán type nonlinear FE models for plates and shells based on classical theory [57], FOSD [58] and TOSD [59–61] hypotheses.

With consideration of strong nonlinear effects, more nonlinear strain-displacement terms are included in the models. This kind of nonlinear theory is usually defined as moderate rotation theory, which was first proposed and developed by Librescu and Schmidt [62], Schmidt and Reddy [63], Schmidt and Weichert [64]. Later, Palmerio et al. [65, 66], Kreja et al. [67] implemented the moderate rotations theory into finite element analysis of composite structures.

2.2.2 Large Rotation Nonlinear Modeling

The von Kármán type nonlinear theory is restricted to weak nonlinearity and small rotations, while the moderate rotation theory is limited to moderately strong nonlinearity and rotations. Both of them are invalid for structures with strong nonlinearity and large rotations. To consider strong nonlinear effects, full geometrically nonlinear strain-displacement relations based on FOSD hypothesis were first developed by Habip [68], Habip and Ebcioglu [69] for static and dynamic equations of shells. Librescu [70] developed fully geometrically nonlinear plate and shell theory for composite laminated structures.

In order to analyze thin-walled structures with large rotations, fully geometrically nonlinear models with finite rotations based on the FOSD hypothesis were applied into FE analysis by Gruttmann et al. [71], Basar et al. [72, 73], Sansour and

Bufler [74], Wriggers and Gruttmann [75], Sansour and Bednarczyk [76], Brank et al. [77], Bischoff and Ramm [78], Kreja and Schmidt [79], Lentzen [80] and others. Kuznetsov and Levyakov [81] developed a fully geometrically nonlinear model with large rotations based on the Kirchhoff-Love theory. Moreover, large rotation nonlinear models were developed for beam or arch structures by Saravia et al. [82], Miller and Palazotto [83].

For relatively thick plates and shells, the TOSD hypothesis was implemented into the large rotation theory by Basar et al. [84, 85] for composite structures, which assumes inextensible shell director yielding seven parameters. Similar TOSD nonlinear models were developed by Bischoff and Ramm [78], Gummadi and Palazotto [86, 87]. Later, Arciniega and Reddy [88] implemented the second-order shear deformation (SOSD) hypothesis into large rotation theory. The SOSD hypothesis assumes a quadratic displacement distribution along the thickness direction. In the model, 3-dimensional constitutive equations was applied, which indicates that the shell director is considered as extensible.

Concerning with soft materials, Basar and Ding [89] developed a nonlinear model considering large strains by taking into account the the transverse normal strain based on SOSD hypothesis. To avoid shear locking phenomenon, large rotation models with four-node assumed strain elements were developed by Dvorkin and Bathe [90], Stander et al. [91], and a nonlinear model with four-node mixed interpolation elements was proposed by Sze et al. [92]. In addition, fully geometrically nonlinear models with using solid elements were developed by Kožar and Ibrahimbegović [93], Masud et al. [94], Lopez and Sala [95] for static analysis of shell structures.

Large or finite rotation theories presented in some publications were not permitting arbitrarily large rotations of the shell director, even though fully geometrically nonlinear strain-displacement relations were considered. Large or finite rotation theories are those which not only consider fully geometrically nonlinear phenomena but also take into account unrestricted rotations. There are two typical approach for large rotation representation, namely Euler angles formulation and Rodrigues rotation formulation, see [96] for the detailed classification. In the FOSD hypothesis, large rotation theory usually includes six independent kinematic parameters. Neglecting the drilling rotation in plates and shells, two rotational variables are proposed to represent last three kinematic parameters. The first approach, Euler angle formulation, was implemented to represent large rotations by Gruttmann et al. [71], Bruechter and Ramm [97], Basar et al. [73], Wriggers and Gruttmann [75], Brank et al. [77], Kreja and Schmidt [79] and others. Additionally, the Rodrigues rotation formulation was proposed by Simo et al. [98, 99]. Later, it was implemented and applied by Sansour and Bufler [74], Betsch et al. [96, 100], Basar et al. [101], Wang and Thierauf [102], Lentzen [80].

2.2.3 Shear Locking Phenomena

Due to the inconsistencies between element representation and transverse shear energy or membrane energy, plate and shell elements may exhibit over stiffening, especially if the thickness tends to be zero. Locking problems are usually referred to shear locking and membrane locking. Shear locking is caused by the Kirchhoff constraints or shear constraints of vanishing transverse shear strains, while membrane locking results from hidden constraints in shell models. The details of introduction of locking phenomena can be found in e.g. [88, 103, 104] among others. To avoid locking problems many numerical methods were proposed and developed e.g. assumed natural strain (ANS) [90, 105–107], enhanced assumed strain (EAS) [108–111], selectively reduced integration (SRI) [112] and uniformly reduced integration (URI) [113–115]. Alternatively, locking effects can be reduced by increasing the number of elements for structures or the number of nodes in an element. Increasing the number of nodes in an element will directly result in higher-order polynomial functions. The method is also known as h-p finite element method, which was proposed and developed earlier by Pitkäranta et al. [103, 116], Leino and Pitkäranta [104] and later by Ref. [88, 117, 118].

2.3 Geometrically Nonlinear Modeling for Smart Structures

Linear models are only valid for smart structures undergoing small displacements and under weak electric fields. When large displacements and rotations occur, geometrically nonlinear theories should be considered in FE models. With consideration of different nonlinear effects and permission of different levels of rotations, various geometrically nonlinear theories were proposed and developed, e.g. von Kármán type nonlinear theory, moderate rotation nonlinear shell theory, fully geometrically nonlinear theory with moderate rotations, and large rotation nonlinear theory. The number of papers dealt with geometrically nonlinear analysis are much less than those with linear analysis.

2.3.1 Von Kármán Type Nonlinear Theory

The von Kármán type nonlinear theory is the simplest nonlinear theory, which is used very frequently in nonlinear analysis of smart structures. The theory contains only the squares and products of derivatives of the transverse deflection in the in-plane longitudinal and shear strain components. The theory is only valid for structures undergoing moderate displacements and small rotations. Im and Atluri [119] first applied von Kármán type nonlinear theory into analysis of piezoelectric integrated structures. Later, von Kármán type nonlinear FE models were developed based on the classical plate theory [120], FOSD hypothesis [121] for buckling analysis of piezoelectric structures. The von Kármán nonlinear FE models with FOSD hypothesis were studied by Panda and Ray [122], Varelis and Saravanos [123], for static analysis of smart structures, and by Mukherjee and Chaudhuri [124] for dynamic analysis.

Implementation of higher-order plate/shell hypothesis, Schmidt and Vu [125] developed von Kármán nonlinear FE models based on both FOSD and TOSD hypotheses for static and dynamic analysis of piezoelectric plates and shells. Cheng et al. [126] carried out a similar study based on TOSD hypothesis for dynamic analysis. Shen and Yang [127], Singh et al. [128] developed nonlinear models using higher-order through-thickness hypothesis.

Considering zigzag hypothesis for laminated structures, von Kármán nonlinear FE models based on the first-order zigzag hypothesis were developed by Carrera [129], Kapuria and Alam [130] for static analysis, and by Ray and Shivakumar [131], Sarangi and Ray [132] for dynamic analysis. Furthermore, Icardi and Sciuva [133] implemented a third-order zigzag hypothesis into geometrically nonlinear analysis of piezoelectric structures.

2.3.2 Moderate Rotation Nonlinear Theory

The von Kármán type nonlinear theory is restricted to moderate rotations and small displacements, since weak geometrical nonlinearity is included in the theory. Considering more nonlinear effects and under the assumption of moderate rotations, a moderate rotation nonlinear shell theory was proposed and develop by Librescu and Schmidt [62], Schmidt and Reddy [63] for composite lamination. The moderate rotation nonlinear theory considers more nonlinear effects, but the strain-displacement terms are still limited. Therefore, the theory is classified into simplified nonlinear theory. Afterwards, the theory was implemented into static and dynamic analysis of smart structures by Lentzen and Schmidt [134], Lentzen et al. [135], Lentzen [80] based on the FOSD hypothesis. Furthermore, for the purpose of comparison, moderate rotation theory with the FOSD hypothesis was investigated by Zhang and Schmidt [136–138], Zhang [23] for static and dynamic analysis of smart structures.

2.3.3 Fully Geometrically Nonlinear Theory with Moderate Rotations

Both von Kármán type nonlinear theory and moderate rotation nonlinear theory consider limited nonlinear effects and under the assumption of small or moderate rotations. To simulate structures with strong nonlinear effects, fully geometrically nonlinear strain-displacement relations should be considered for smart structures. If the assumption of moderate rotations is still imposed, the resulting theory is fully geometrically nonlinear theory with moderate rotations. Due to the kinematic hypothesis of moderate rotations, the results obtained by fully geometrically nonlinear theory with moderate rotations are close to those obtained by moderate rotation nonlinear theory.

Based on the Kirchhoff-Love hypothesis, Moita et al. [139] developed a fully geometrically nonlinear FE model for static analysis of smart structures. Based on the FOSD hypothesis, fully geometrically nonlinear FE models were developed by Kundu et al. [140] for buckling and post-buckling analysis, by Gao and Shen [114] for dynamic analysis. Implementation of the TOSD hypothesis into fully geometrically nonlinear FE model, Dash and Singh [141] studied for dynamic analysis.

Considering geometrical imperfections in the thickness direction, fully geometrically nonlinear FE models were developed by Amabili [142] based on the FOSD hypothesis, and by Amabili [143] based on the TOSD hypothesis. Additionally, based on the higher-order shear deformation hypothesis, Alijani and Amabili [144, 145] built fully geometrically nonlinear FE models with consideration of thickness stretching. Amabili [146], Amabili and Reddy [147] included both geometrical imperfection and thickness stretching in the fully geometrically nonlinear FE models for composite structures.

2.3.4 Large Rotation Nonlinear Theory

The nonlinear theories including von Kármán type nonlinear theory, moderate rotation nonlinear theory, and fully geometrically nonlinear theory with moderate rotations, are only applicable to structures undergoing large displacements and moderate rotations. Due to this limitations, the theories invalid for the structures undergoing large displacements and rotations, which are thus classified as simplified nonlinear theories.

Considering fully geometrically nonlinear strain-displacement relations with large rotations yields large rotation nonlinear theory. Chróścielewski et al. [148–150] developed a 1D FE model of large rotation nonlinear theory for shape and vibration control of arches. Zhang and Schmidt [136–138, 151] proposed a large rotation nonlinear FE model with the FOSD hypothesis for static and dynamic analysis of piezolaminated plate and shell structures. The unrestricted rotations are updated by using Euler rotation formulation. Analogously, Rao and Schmidt [152], Rao et al. [153] studied a similar large rotation nonlinear model by using Rodrigues rotation formulation.

2.4 Electroelastic Materially Nonlinear Modeling

2.4.1 Linear Piezoelectric Constitutive Equations

In most of the studies, linear constitutive laws were employed in finite element models of smart structures, which are only valid for structures under weak electric field. There are two typical models of electric potential through the thickness, namely first-order and higher-order variation. The former distribution of electric potential is mostly used in modeling of piezoelectric materials, which yields constant electric field through the thickness. Because of the assumption of linear variation of electric potential, it is applicable only for thin piezoelectric patches. Almost all the above mentioned models of smart structures implemented this variation of electric potential.

In the case of thick piezoelectric layers, higher-order variation of electric potential should be considered [154, 155]. Linear models with quadratic electric potential variation through the thickness were developed based on the FOSD hypothesis [28, 156] and zigzag hypothesis [157]. Using the MITC elements, proposed by Dvorkin and Bather [90], Bathe [105], FE models with the assumption of second-order variation of electric potential were proposed by Kögl and Bucalem [158]. Moreover, geometrically nonlinear FE models with electric potential quadratic distribution were developed for static and dynamic analysis [159, 160].

2.4.2 Strong Electric Field Models

Linear piezoelectric constitutive equations are only used when the structures undergo small strains and under weak electric potential. In piezoelectric material, it is assumed that the stresses generated by electric field is always below the yield stress, meaning that structures undergo only in small strains. However, sometimes strong electric field is considered to be applied on piezoelectric material for large actuation forces. This requires an electroelastic materially nonlinear relations. Therefore, for the case of small strains and strong electric field, the nonlinear part of constitutive law includes only the electroelastic part.

The constitutive equations with electroelastic nonlinearity were first proposed by Nelson [161] and Joshi [162]. Afterwards, the constitutive equations were extended and implemented into transversely isotropic materials like piezoelectric ceramics and the class of mm^2 symmetry materials like PVDF [163]. Many researchers investigated irreversible piezoelectric nonlinearities, known as piezoelectric hysteresis, e.g. [164–168] among many others. To validate the numerical models of piezoelectric hysteresis, Li et al. [169], Masys et al. [170] investigated experimentally. In addition, Klinkel [171], Linnemann et al. [172] applied the irreversible phenomenological constitutive model into finite element analysis using solid elements for piezoelectric materials.

For beam structures, Tan and Tong [173] studied a one-dimensional analytical model with consideration of electroelastic nonlinear effect of piezoelectric fiber reinforced composite materials through the curve-fitting method based on experimental data. Wang et al. [174] developed electroelastic nonlinear analytical models for clamped piezoelectric bimorph and unimorph beams, with experimentally validated. Analogously, Yao et al. [175] developed a very similar nonlinear beam model with electroelastic nonlinearity and tested by experimental investigations on bimorph and unimorph beams.

Regarding to plates and shells integrated with piezoelectric materials, many papers can be found in the literature that developed electroelastic materially nonlinear numerical models. Sun et al. [176], Kusculuoglu and Royston [177] developed finite element models with electroelastic material nonlinearity based on Reissner-Mindlin plate hypothesis for static shape control and dynamic analysis of smart structures. Kapuria and Yasin [178, 179] proposed nonlinear FE models based on layerwise theory for the static analysis and active vibration control of piezoelectric structure under strong electric field. Using the model of quadratic distribution of electric potential, Rao et al. [180] proposed an FE model with consideration of electroelastic materially nonlinear effects for piezoelectric laminated composite plates and shells.

The above mentioned studies in this subsection are mainly focusing on geometrically linear models with electroelastic materially nonlinear effect, which allows structures only undergoing small displacements. When structures undergo large displacements and under strong electric fields, both geometrically and electroelastic materially nonlinear effects should be included in the numerical models. Yao et al. [181] developed a nonlinear model with von Kármán type nonlinearity based on the classical plate theory for structures under strong driving electric field. Zhang et al. [182] proposed a fully nonlinear model with both geometrically nonlinear (large rotation nonlinear) and electroelastic materially nonlinear effects for piezolaminated smart structures.

2.5 Multi-physics Coupled Modeling

2.5.1 Functionally Graded Structures

Smart structures consist of conventional piezoelectric and metal materials. With the development of material science, many advanced materials were invented, like carbon nanotube (CNT) reinforced functionally graded composites, functionally graded piezoelectric materials. Piezoelectric smart structures are inherently coupled with electro-mechanical fields. On one hand, multi-physics coupled modeling techniques are necessary for precise structural computation, on the other hand, modeling of new material structures should be developed.

Carbon nanotube reinforced functionally graded composites bonded with piezoelectric layers have excellent mechanical and electrical performance, attracting many researchers. A linear model of functionally graded CNT reinforced composite cylindrical shell layered with piezoelectric materials was proposed by Alibeigloo [183] for vibration analysis. An element-free model based on Reddy's higher-order shear deformation hypothesis was developed by Selim et al. [184] for CNT reinforced composite plate boned with piezo layers.

Considering structures with large deformation, a von Kármán type geometrically nonlinear model based on the classical plate theory was developed by Rafiee et al. [185] for CNT reinforced functionally graded composite beams with piezoelectric materials. Including additionally the thermal effect, a von Kármán type nonlinear analytical model was proposed by Ansari et al. [186] for postbuckling analysis of functionally graded CNT reinforced composite cylindrical shells under electro-thermal hybrid loading conditions.

Applying functionally graded concept to piezoelectric materials, one obtains functionally graded piezoelectric material. Loja et al. [187] developed B-spline finite strip element models for sandwich structures with functionally graded piezoelectric materials. Mikaeeli and Behjat [188] investigated static analysis of thick functionally graded piezoelectric plates using three dimensional element-free Galerkin method. A finite element model based on the FOSD hypothesis was developed by Su et al. [189] for free vibration and transient analysis of functionally graded piezoelectric plates. For structures undergoing large displacements, Derayatifar et al. [190], Wang [191] developed von Kármán type geometrically nonlinear models for functionally graded piezoelectric material integrated smart structures.

2.5.2 *Electro-Thermo-Mechanically Coupled Structures*

For every structure, all the materials are exposed to thermal field. Many of them are sensitive to the change of thermal field. Considering the electro-thermo-mechanically coupled structures, Krommer and Irschik [192] proposed a finite element model based on the Reissner-Mindlin theory. Zhang et al. [193], Li et al. [194] developed thermo-electro-mechanically coupled models based on the classical plate theory for analysis of piezoelectric nanoplates with viscoelastics. Arefi and Zenkour [195] investigated thermo-electro-mechanical bending behavior of sandwich nanoplates integrated with piezoelectric face-sheets using trigonometric plate theory. In addition, three-dimensional equations coupled with electro-thermo-mechanical fields were studied by Dehghan et al. [196] for functionally graded piezoelectric shells.

Thermal analysis was investigated on CNT reinforced functionally graded composites by many researchers. An analytical solution of thermal coupled analysis was proposed by Alibeigloo [197]. A linear model based on the Reddy's higher-order shear deformation hypothesis was developed by Song et al. [198]. Considering material nonlinearity, an electro-thermo-elasto-plastic model was develop by Tang and Felicelli [199] using an incremental formulation based on the variational-asymptotic method.

Including geometrical nonlinearity, von Kármán type nonlinear models coupled with thermal effects were developed based on Timoshenko beam theory [200] and Reddy's higher-order theory[127]. A Sanders nonlinear model based on classical shell theory was proposed for snap-through buckling analysis of functionally graded structures with thermally coupled [201].

2.5.3 *Magneto-Electro-Elastic Composites*

Magneto-electro-elastic structures are also known as MEE structures, which couple with electric, magnetic and elastic fields. MEE structures are capable of energy conversion among the forms of magnetic, electric and elastic. Vinyas and Kattimani [202] developed a 3D FE model for hygrothermal analysis of MEE plate, while Yang et al. [203] studied a similar model for natural characteristic analysis. Additionally, numerical models based on Donnell theory [204] or a four-variable shear deformation refined plate theory [205] were developed for MEE plates.

In the application of piezoelectric-piezomagnetic functionally graded materials with a gradual change of the mechanical and electromagnetic properties, Ezzin et al. [206] proposed a dynamic solution based on the ordinary differential equation and stiffness matrix methods for the propagation of waves on a structure covered with a functionally graded piezoelectric material layer. For structures undergoing large displacements, von Kármán type nonlinear models were developed based on the FOSD hypothesis [207] and first-order zigzag hypothesis [208] for MEE sandwich plate. Furthermore, a geometrically nonlinear model, a nonlocal strain gradient shell model, was developed for buckling and postbuckling analysis of MEE composites [209].

2.5.4 *Aero-Electro-Elastic Coupled Modeling*

One of the most important applications of smart structures is flutter control of aircraft panels, in which fluid-solid interaction is the basic feature of the problems. Taking into account fluid, electric and elastic coupled fields, Wang et al. [210], Song and Li [211], Li [212] developed linear aero-electro-elastic models of piezoelectric plates for flutter suppression under supersonic air flows. Considering more physical field, like thermal field, Mohammadimehr and Mehrabi [213], Song et al. [214] developed aero-electro-thermo-elastic coupled FE models for vibration and flutter analysis of supersonic piezoelectric composite plate.

2.6 Modeling of Piezo-Fiber Composite Bonded Structures

2.6.1 Types of Piezo Fiber Composite Materials

Is is known that piezoelectric ceramics are brittle and piezoelectric polymers are with weak actuation forces. To overcome the limitations of conventional piezoelectric materials, piezoelectric fiber based composites were invented through mixture of piezoceramic fibrous phase and epoxy matrix phase.

The first type of piezo-fiber based composite material was proposed by Skinner et al. [215], known as 1-3 composite. In the type of 1-3 composite, the piezoelectric fibers with rectangular or circular cross section place along in the thickness direction. Due to the piezoelectric fiber orientation, this type of piezo composite still has weak actuation forces along the in-plane directions.

Placing the piezoelectric fiber with circular cross section along the in-plane direction, one obtains an active fiber composite (AFC), initially invented by MIT [216, 217]. Because of the circular cross section, a certain electric field volume is invalid for the actuation performance. Replacing the circular cross section with rectangular cross section, yields a macro-fiber composite (MFC), which was invented by NASA Langley Research Center [218]. The MFC piezoelectric composites have no loss on electric field, resulting in large actuation forces. For more details of MFC piezoelectric composites, it refers to Williams et al. [219], Sodano et al. [220], Bowen et al. [221]. Since MFC has many beneficial properties, many applications for vibration control [222, 223] and health monitoring [224–226] were investigated.

2.6.2 Homogenization of Piezo Fiber Composite

The structures of fiber based piezoelectric composite are complicated. For easy implementation in simulations, piezoelectric composites are usually homogenized to orthotropic materials, by experimental and numerical investigations. MFCs have large application potentials due to their beneficial properties. Therefore, most of the studies were dealing with the homogenization of MFC materials. Williams et al. [227], Williams [228] obtained the basic elasticity constants of MFC patches for the elastic and plastic constitutive behavior through experimental investigations. Linear piezoelectric composite material properties were predicted by using classical lamination theory [229], representative volume element (RVE) technique with mixing rules [230–232], and asymptotic expansion homogenization (AEH) method [233]. More precisely, an electroelastic nonlinear material constitutive equations was developed by Williams et al. [234] for MFC patches. In addition, hysteresis and creep effects were studies experimentally by Schröck et al. [235] for dynamic performance of MFC integrated structures. For achieving complete material parameters, including not only the elastic constants but also the transverse shear moduli and the piezoelectric constants, Li et al. [236], Trindade and Benjeddou [237] proposed homogenization approaches for MFC patches.

2.6.3 *Modeling of Piezo Composite Laminated Plates and Shells*

To investigate the structural response, the earlier work on simulation of piezo composite laminated plates and shells is mainly using commercial software, e.g. ANSYS [238, 239], ABAQUS [240, 241], which was validated by the experimental results. The numerical studies of snap-through of asymmetric bistable curved laminates with MFC were performed by Bowen et al. [239] and Giddings et al. [242]. Using the commercial software, Bilgen et al. [243] developed a linear distributed parameter electro-mechanical model for dynamic analysis of MFC bonded cantilevered thin beams. To study the influence of piezo-fiber orientation, Zhang et al. [244, 245] developed a linear FE model of MFC embedded thin-walled structures based on the Reissner-Mindlin hypothesis for both static and dynamic analysis with variation of piezo fiber orientation angle.

Considering geometrically nonlinear phenomenon in the simulation, Azzouz and Hall [246] proposed a nonlinear FE model with a von Kármán type nonlinearity based on the Reissner-Mindlin hypothesis for dynamic analysis of a rotating MFC bonded plates. Moreover, Zhang et al. [247] developed various geometrically nonlinear finite element models based on the FOSD hypothesis using e.g. von Kármán type nonlinear theory, moderate rotation nonlinear theory, fully geometrically nonlinear theory with moderate rotations and large rotations nonlinear theory for static analysis of MFC bonded plates and shells.

2.7 Vibration Control of Piezo Smart Structures

2.7.1 *Conventional Control Strategies*

Smart structures have a great potential in the field of vibration control. On one hand, the design of smart structure has great impact on the efficiency of vibration control, on the other hand, the design of control law are of equal importance. By literature review, it revels that most studies were developed conventional control laws based on linear FE models.

The most frequently used control law is negative velocity proportional feedback control. A lot of publications have been implemented it into vibration control of smart structures, using linear FE models based on various hypotheses, see [19–21, 30–32, 35, 41, 45, 54, 248–262]. Moreover, Moita et al. [41] studied optimization of piezoelectric position for negative velocity proportional feedback control using genetic algorithm. Besides, the negative velocity proportional feedback control law, Lyapunov feedback control [30, 54, 248–250] and bang-bang control [52, 223] were also investigated by many researchers.

Many studies were implemented optimal control laws into simulations of vibration control of smart structures. Linear quadratic regulator (LQR) control is a full state feedback control, investigated by Kang et al. [253], Narayanan and Balamurugan [54], Balamurugan and Narayanan [30], Raja et al. [263], Vasques and Rodrigues [261], Valliappan and Qi [264] and Xu and Koko [265]. LQR control is an ideal method, which assumes that all the state variables should be measurable and fed back to the controller. However, state variables can not be all measured in real applications. Therefore, linear quadratic Gaussian (LQG) control was applied to smart structures by Stavroulakis et al. [266], Vasques and Rodrigues [261], Dong et al. [267]. In LQG control, the state variables are not necessarily measured, but can be estimated by an observer. Furthermore, Marinaki et al. [55] proposed a particle swarm optimization based controller for vibration suppression of beams. Roy and Chakraborty [268] developed a genetic algorithm based LQR control for smart composite shell structures.

2.7.2 Advanced Control Strategies

Conventional controls are with easy implementation, but they have low control efficiency and robustness. To improve the control effect, Chen and Shen [269], Lin and Nien [270] developed an independent modal space control for vibration suppression of smart structures. Bhattacharya et al. [271] proposed an independent modal space based LQR control strategy for vibration control of laminated spherical shell with various fiber orientation and curvature radius. Furthermore, Manjunath and Bandyopadhyay [272] developed a discrete sliding mode control scheme, Valliappan and Qi [264] proposed a prediction control algorithm for smart beams with bonded piezoelectric patches. Zhang et al. developed disturbance rejection control with both proportional-integral (PI) [273, 274] and generalized-proportional-integral (GPI) observers [274] for vibration suppression of smart structures. Later, in the framework of disturbance rejection control, Zhang et al. [275], Zhang et al. [276] developed generalized disturbance rejection control with PI observer for smart beams.

Considering finite element models with geometric nonlinearities, very less papers can be found in the literature dealing with control simulations. Due to the complexity of nonlinear numerical models, most of the studies were applying very simple control schemes, Zhou and Wang [277] applied a negative velocity or displacement feedback control for vibration suppression of beams. In addition, Schmidt and Vu [125], Vu [278], Lentzen and Schmidt [134, 135] investigated the same control schemes for vibration suppression of piezoelectric bonded plate structures based on von Kármán type nonlinear FE models, while Gao and Shen [114] studied based on a fully geometrically nonlinear FE plate model with FOSD hypothesis.

2.7.3 *Intelligent Control Strategies*

Most of the control schemes need accurate mathematical models for control design. If the mathematical models are difficult to construct, intelligent control is the best approach for vibration suppression, e.g. neural network control, fuzzy logic control. Intelligent control has been develop in the last several decades for various applications. However, only a very limited publications implemented intelligent control into vibration suppression of smart structures. Lee [279], Han and Acar [280], Valoor et al. [281] developed neural network control for simulation of vibration suppression of smart structures, Youn et al. [282], Kumar et al. [283], Qiu et al. [284] applied into experimental investigations. In addition, Jha and He [285] developed a neural adaptive predictive control for smart structures. Shirazi et al. [286], Abreu and Ribeiro [287] proposed a fuzzy logic control for vibration suppression of functionally graded rectangular plate integrated with piezoelectric patches.

References

1. H.S. Tzou, C.I. Tseng, Distributed piezoelectric sensor/actuator design for dynamic measurement/control of distributed parameter systems: a piezoelectric finite element approach. J. Sound Vib. **138**, 17–34 (1990)
2. S.K. Ha, C. Keilers, F.K. Chang, Finite element analysis of composite structures containing distributed piezoceramic sensors and actuators. AAIA J. **30**(3), 772–280 (1992)
3. G.P. Dube, S. Kapuria, P.C. Dumir, Exact piezothermoelastic solution of simply-supported orthotropic flat panel in cylindrical bending. Int. J. Mech. Sci. **38**, 1161–1177 (1996)
4. S. Kapuria, G.P. Dube, Exact piezothermoelastic solution for simply supported laminated flat panel in cylindrical bending. ZAMM · Z. Angew. Math. Mech. **77**, 281–293 (1997)
5. M.C. Ray, R. Bhattacharya, B. Samanta, Exact solutions for dynamic analysis of composite plates with distributed piezoelectric layers. Comput. Struct. **66**, 737–743 (1998)
6. L.H. He, Three dimensional analysis of some symmetric hybrid piezoelectric laminates. ZAMM · Z. Angew. Math. Mech. **80**, 307–318 (2000)
7. K.Y. Sze, L.Q. Yao, S. Yi, A hybrid stress ANS solid-shell element and its generalization for smart structure modeling: part II smart structure modeling. Int. J. Numer. Methods Eng. **48**, 565–582 (2000)
8. K.Y. Sze, L.Q. Yao, A hybrid stress ANS solid-shell element and its generalization for smart structure modeling: part I solid shell element formulation. Int. J. Numer. Methods Eng. **48**, 545–564 (2000)
9. K.Y. Sze, L.Q. Yao, Modeling smart structures with segmented piezoelectric sensors and actuators. J. Sound Vib. **235**, 495–520 (2000)
10. S. Kapuria, P. Kumari, Three-dimensional piezoelasticity solution for dynamics of cross-ply cylindrical shells integrated with piezoelectric fiber reinforced composite actuators and sensors. Compos. Struct. **92**, 2431–2444 (2010)
11. S. Yi, S.F. Ling, M. Ying, Large deformation finite element analyses of composite structures integrated with piezoelectric sensors and actuators. Finite Elem. Anal. Des. **35**, 1–15 (2000)
12. S. Klinkel, W. Wagner, A geometrically non-linear piezoelectric solid shell element based on a mixed multi-field variational formulation. Int. J. Numer. Methods Eng. **65**, 349–382 (2006)
13. S. Klinkel, W. Wagner, A piezoelectric solid shell element based on a mixed variational formulation for geometrically linear and nonlinear applications. Comput. Struct. **86**, 38–46 (2008)

14. H.S. Tzou, M. Gadre, Theoretical analysis of a multi-layered thin shell coupled with piezoelectric shell actuators for distributed vibration controls. J. Sound Vib. **132**, 433–450 (1989)
15. C.K. Lee, Theory of laminated piezoelectric plates for the design of distributed sensors/actuators. Part I: governing equations and reciprocal relationships. J. Acoust. Soc. Am. **87**, 1144–1158 (1990)
16. H. Kioua, S. Mirza, Piezoelectric induced bending and twisting of laminated composite shallow shells. Smart Mater. Struct. **9**, 476–484 (2000)
17. K.Y. Lam, X.Q. Peng, G.R. Liu, J.N. Reddy, A finite-element model for piezoelectric composite laminates. Smart Mater. Struct. **6**, 583–591 (1997)
18. D.A. Saravanos, Mixed laminate theory and finite element for smart piezoelectric composite shell structures. AIAA J. **35**, 1327–1333 (1997)
19. G.R. Liu, X.Q. Peng, K.Y. Lam, J. Tani, Vibration control simulation of laminated composite plates with integrated piezoelectrics. J. Sound Vib. **220**, 827–846 (1999)
20. J.M.S. Moita, I.F.P. Correia, C.M. Soares, C.A.M. Soares, Active control of adaptive laminated structures with bonded piezoelectric sensors and actuators. Comput. Struct. **82**, 1349–1358 (2004)
21. J.M.S. Moita, V.M.F. Correia, P.G. Martins, C.M.M. Soares, C.A.M. Soares, Optimal design in vibration control of adaptive structures using a simulated annealing algorithm. Compos. Struct. **75**, 79–87 (2006)
22. R.P. Shimpi, H.G. Patel, Free vibration of plate using two variable refined plate theory. J. Sound Vib. **296**, 979–999 (2006)
23. S.Q. Zhang, *Nonlinear FE Simulation and Active Vibration Control of Piezoelectric Laminated Thin-Walled Smart Structures*. PhD thesis, RWTH Aachen University (2014)
24. R.D. Mindlin, Forced thickness-shear and flexural vibrations of piezoelectric crystal plates. J. Appl. Phys. **23**, 83–88 (1952)
25. C.C. Lin, C.Y. Hsu, H.N. Huang, Finite element analysis on deflection control of plates with piezoelectric actuators. Compos. Struct. **35**, 423–433 (1996)
26. S. Cen, A.K. Soh, Y.Q. Long, Z.H. Yao, A new 4-node quadrilateral FE model with variable electrical degrees of freedom for the analysis of piezoelectric laminated composite plates. Compos. Struct. **58**, 583–599 (2002)
27. S. Kapuria, P.C. Dumir, First order shear deformation theory for hybrid cylindrical panel in cylindrical bending considering electrothermomechanical coupling effects. ZAMM · Z. Angew. Math. Mech. **82**, 461–471 (2002)
28. D. Marinković, H. Köppe, U. Gabbert, Accurate modeling of the electric field within piezoelectric layers for active composite structures. J. Intell. Mater. Syst. Struct. **18**, 503–513 (2007)
29. C.Q. Chen, X.M. Wang, Y.P. Shen, Finite element approach of vibration control using self-sensing piezoelectric actuators. Comput. Struct. **60**, 505–512 (1996)
30. V. Balamurugan, S. Narayanan, Shell finite element for smart piezoelectric composite plate/shell structures and its application to the study of active vibration control. Finite Elem. Anal. Design **37**, 713–738 (2001)
31. S.Y. Wang, S.T. Quek, K.K. Ang, Vibration control of smart piezoelectric composite plates. Smart Mater. Struct. **10**, 637–644 (2001)
32. S.Y. Wang, S.T. Quek, K.K. Ang, Dynamic stability analysis of finite element modeling of piezoelectric composite plates. Int. J. Solids Struct. **41**, 745–764 (2004)
33. M. Krommer, Piezoelastic vibrations of composite Reissner-Mindlin-type plates. J. Sound Vib. **263**, 871–891 (2003)
34. S.Y. Wang, A finite element model for the static and dynamic analysis of a piezoelectric bimorph. Int. J. Solids Struct. **41**, 4075–4096 (2004)
35. G.R. Liu, K.Y. Dai, K.M. Lim, Static and vibration control of composite laminates integrated with piezoelectric sensors and actuators using the radial point interpolation method. Smart Mater. Struct. **13**, 1438–1447 (2004)
36. J.N. Reddy, A simple higher-order theory for laminated composite plates. J. Appl. Mech. **51**, 745–752 (1984)

37. J.N. Reddy, A refined nonlinear theory of plates with transverse shear deformation. Int. J. Solids Struct. **20**, 881–896 (1984)
38. N.F. Hanna, A.W. Leissa, A higher order shear deformation theory for the vibration of thick plates. J. Sound Vib. **170**, 545–555 (1994)
39. V.M.F. Correia, M.A.A. Gomes, A. Suleman, C.M.M. Soares, C.A.M. Soares, Modelling and design of adaptive composite structures. Comput. Methods Appl. Mech. Eng. **185**(2), 325–346 (2000)
40. I.F.P. Correia, C.M.M. Soares, C.A.M. Soares, J. Herskovits, Active control of axisymmetric shells with piezoelectric layers: a mixed laminated theory with a high order displacement field. Comput. Struct. **80**, 2265–2275 (2002)
41. J.S. Moita, P.G. Martins, C.M.M. Soares, C.A.M. Soares, Optimal dynamic control of laminated adaptive structures using a higher order model and a genetic algorithm. Comput. Struct. **86**, 198–206 (2008)
42. B.A. Selim, L.W. Zhang, K.M. Liew, Active vibration control of fgm plates with piezoelectric layers based on reddy's higher-order shear deformation theory. Compos. Struct. **155**, 118–134 (2016)
43. M.A.R. Loja, C.M.M. Soares, C.A.M. Soares, Higher-order B-spline finite strip model for laminated adaptive structures. Compos. Struct. **52**, 419–427 (2001)
44. C.M.M. Soares, C.A.M. Soares, V.M.F. Correia, M.A.R. Loja, Higher-order B-spline strip models for laminated composite structures with integrated sensors and actuators. Compos. Struct. **54**, 267–274 (2001)
45. M.C. Ray, J.N. Reddy, Active control of laminated cylindrical shells using piezoelectric fiber reinforced composites. Compos. Sci. Technol. **65**, 1226–1236 (2005)
46. C.M.A. Vasques, J.D. Rodrigues, Coupled three-layered analysis of smart piezoelectric beams with different electric boundary conditions. Int. J. Numer. Methods Eng. **62**, 1488–1518 (2005)
47. S. Kapuria, An efficient coupled theory for multilayered beams with embedded piezoelectric sensory and active layers. Int. J. Solids Struct. **38**, 9179–9199 (2001)
48. S. Kapuria, P.C. Dumir, A. Ahmed, An efficient coupled layerwise theory for dynamic analysis of piezoelectric composite beams. J. Sound Vib. **261**, 927–944 (2003)
49. O. Polit, M. D'Ottavio, P. Vidal, High-order plate finite elements for smart structure analysis. Compos. Struct. **151**, 81–90 (2016)
50. E. Carrera, L. Demasi, Classical and advanced multilayered plate elements based upon PVD and RMVT. Part 2: numerical implementations. Int. J. Numer. Methods Eng. **55**, 253–291 (2002)
51. E.F. Crawley, J. Luis, Use of piezoelectric actuators as elements of intelligent structures. AIAA J. **25**(10), 1373–1385 (1987)
52. H.S. Tzou, W.K. Chai, Design and testing of a hybrid polymeric electrostrictive/piezoelectric beam with bang-bang control. Mech. Syst. Signal Process. **21**, 417–429 (2007)
53. I. Kucuk, I.S. Sadek, E. Zeihi, S. Adali, Optimal vibration control of piezolaminated smart beams by the maximum principle. Comput. Struct. **89**, 744–749 (2011)
54. S. Narayanan, V. Balamurugan, Finite element modeling of piezolaminated smart structures for active vibration control with distributed sensors and actuators. J. Sound Vib. **262**, 529–562 (2003)
55. M. Marinaki, Y. Marinakis, G.E. Stavroulakis, Vibration control of beams with piezoelectric sensors and actuators using partical swarm optimization. Expert Syst. Appl. **38**, 6872–6883 (2011)
56. Y. Zhu, J.W. Zu, Modeling of piezoelectric energy harvester: a comparison between Euller-Bernoulli and Timoshenko theory, in *Proceedings of the ASME 2011 Conference on Smart Materials, Adaptive Structures and Intelligent Systems*, Phoenix, Arizona, USA, September 18–21 (2011)
57. P. Ribeiro, Non-linear free periodic vibrations of open cylindrical shallow shells. J. Sound Vib. **313**, 224–245 (2008)
58. J.N. Reddy, Geometrically nonlinear transient analysis of laminated composite plates. AIAA J. **21**, 621–629 (1983)

59. J.N. Reddy, On refined computational models of composite laminates. Int. J. Numer. Methods Eng. **27**, 361–382 (1989)
60. J.N. Reddy, A general non-linear third-order theory of plates with moderate thickness. Int. J. Non-Linear Mech. **25**, 677–686 (1990)
61. M.E. Fares, M.Kh. Elmarghany, A refined zigzag nonlinear first-order shear deformation theory of composite laminated plates. Compos. Struct. **82**, 71–83 (2008)
62. L. Librescu, R. Schmidt, Refined theories of elastic anisotropic shells accounting for small strains and moderate rotations. Int. J. Non-Linear Mech. **23**, 217–229 (1988)
63. R. Schmidt, J.N. Reddy, A refined small strain and moderate rotation theory of elastic anisotropic shells. J. Appl. Mech. **55**, 611–617 (1988)
64. R. Schmidt, D. Weichert, A refined theory of elastic-plastic shells at moderate rotations. ZAMM · Z. Angew. Math. Mech. **69**, 11–21 (1989)
65. A.F. Palmerio, J.N. Reddy, R. Schmidt, On a moderate rotation theory of laminated anisotropic shells - part 1: theory. Int. J. Non-Linear Mech. **25**, 687–700 (1990)
66. A.F. Palmerio, J.N. Reddy, R. Schmidt, On a moderate rotation theory of laminated anisotropic shells - part 2: finite element analysis. Int. J. Non-Linear Mech. **25**, 701–714 (1990)
67. I. Kreja, R. Schmidt, J.N. Reddy, Finite elements based on a first-order shear deformation moderate rotation shell theory with applications to the analysis of composite structures. Int. J. Non-Linear Mech. **32**, 1123–1142 (1996)
68. L.M. Habip, Theory of elastic shells in the reference state. Ingenieur-Archiv **34**, 228–237 (1965)
69. L.M. Habip, I.K. Ebcioglu, On the equations of motion of shells in the reference state. Ingenieur-Archiv **34**, 28–32 (1965)
70. L. Librescu, *Elastostatics and Kinetics of Anisotropic and Heterogeneous Shell-Type Structures* (Noordhoff International, Leyden, 1975)
71. F. Gruttmann, E. Stein, P. Wriggers, Theory and numerics of thin elastic shells with finite rotations. Ingenieur-Archiv **59**, 54–67 (1989)
72. Y. Basar, Y. Ding, Finite-rotation shell elements for the analysis of finite-rotation shell problems. Int. J. Numer. Methods Eng. **34**, 165–169 (1992)
73. Y. Basar, Y. Ding, W.B. Krätzig, Finite-rotation shell elements via mixed formulation. Comput. Mech. **10**, 289–306 (1992)
74. C. Sansour, H. Bufler, An exact finite rotation shell theory, its mixed variational formulation and its finite element implementation. Int. J. Numer. Methods Eng. **34**, 73–115 (1992)
75. P. Wriggers, F. Gruttmann, Thin shells with finite rotations formulated in biot stresses: theory and finite element formulation. Int. J. Numer. Methods Eng. **36**, 2049–2071 (1993)
76. C. Sansour, H. Bednarczyk, The cosserat surface as a shell model, theory and finite-element formulation. Comput. Methods Appl. Mech. Eng. **120**, 1–32 (1995)
77. B. Brank, D. Perić, B. Damjanić, On implementation of a nonlinear four node shell finite element for thin multilayered elastic shells. Comput. Mech. **16**, 341–359 (1995)
78. M. Bischoff, E. Ramm, Shear deformable shell elements for large strains and rotations. Int. J. Numer. Methods Eng. **40**, 4427–4449 (1997)
79. I. Kreja, R. Schmidt, Large rotations in first-order shear deformation FE analysis of laminated shells. Int. J. Non-Linear Mech. **41**, 101–123 (2006)
80. S. Lentzen, *Nonlinear coupled thermopiezoelectric modelling and FE-simulation of smart structures*. Ph.D. Thesis, RWTH Aachen University, in: Fortschritt-Berichte VDI, Reihe 20, Nr. 419, VDI Verlag, Düsseldorf (2009)
81. V.V. Kuznetsov, S.V. Levyakov, Geometrically nonlinear shell finite element based on the geometry of a planar curve. Finite Elem. Anal. Des. **44**, 450–461 (2008)
82. C.M. Saravia, S.P. Machado, V.H. Cortinez, A geometrically exact nonlinear finite element for composite closed section thin-walled beams. Comput. Struct. **89**, 2337–2351 (2011)
83. D.A. Miller, A.N. Palazotto, Nonlinear finite element analysis of composite beams and arches using a large rotation theory. Finite Elem. Anal. Des. **19**, 131–152 (1995)
84. Y. Basar, Y. Ding, R. Schultz, Refined shear-deformation models for composite laminates with finite rotations. Int. J. Solids Struct. **30**, 2611–2638 (1993)

85. Y. Basar, Finite-rotation theories for composite laminates. Acta Mechanica **98**, 159–176 (1993)
86. L.N.B. Gummadi, A.N. Palazotto, Finite element analysis of arches undergoing large rotations - I: theoretical comparison. Finite Elem. Anal. Des. **24**, 213–235 (1997)
87. A.N. Palazotto, L.N.B. Gummadi, J.C. Bailey, Finite element analysis of arches undergoing large rotations - II: classification. Finite Elem. Anal. Des. **24**, 237252 (1997)
88. R.A. Arciniega, J.N. Reddy, Tensor-based finite element formulation for geometrically nonlinear analysis of shell structures. Comput. Methods Appl. Mech. Eng. **196**, 1048–1073 (2007)
89. Y. Basar, Y. Ding, Shear deformation models for large-strain shell analysis. Int. J. Solids Struct. **14**, 1687–1708 (1997)
90. E.N. Dvorkin, K.J. Bather, A continuum mechanics based four-node shell element for general non-linear analysis. Eng. Comput. **1**, 77–88 (1984)
91. N. Stander, A. Matzenmiller, E. Ramm, An assessment of assumed strain methods in finite rotation shell analysis. Eng. Comput. **6**, 58–66 (1989)
92. K.Y. Sze, X.H. Liu, S.H. Lo, Popular benchmark problems for geometric nonlinear analysis of shells. Finite Elem. Anal. Des. **40**, 1551–1569 (2004)
93. I. Kožar, A. Ibrahimbegović, Finite elemtent formulation of the finite rotation solid element. Finite Elem. Anal. Des. **20**, 101–126 (1995)
94. A. Masud, C.L. Tham, W.K. Liu, A stabilized 3-D co-rotational formulation for geometrically nonlinear analysis of multi-layered composite shells. Comput. Mech. **26**, 1–12 (2000)
95. S. Lopez, G.L. Sala, A finite element approach to statical and dynamical analysis of geometrically nonlinear structures. Finite Elem. Anal. Des. **46**, 1093–1105 (2010)
96. P. Betsch, A. Menzel, E. Stein, On the parametrization of finite rotations in computational mechanics-a classification of concepts with application to smooth shells. Comput. Methods Appl. Mech. Eng. **155**, 273–305 (1998)
97. N. Bruechter, E. Ramm, Shell theory versus degeneration-a comparison in large rotation finite element analysis. Int. J. Numer. Methods Eng. **34**, 39–59 (1992)
98. J.C. Simo, M.S. Rifai, D.D. Fox, On a stress resultant geometrically exact shell model. Part VI: conserving algorithms for non-linear dynamics. Int. J. Numer. Methods Eng. **34**, 117–164 (1992)
99. J.C. Simo, On a stress resultant geometrically exact shell model. Part VII: shell intersections with 5/6 DOF finite element formulations. Comput. Methods Appl. Mech. Eng. **108**, 319–339 (1993)
100. P. Betsch, F. Gruttmann, E. Stein, A 4-node finite shell element for the implementation of general hyperelastic 3D-elasticity at finite strains. Comput. Methods Appl. Mech. Eng. **130**, 57–79 (1996)
101. Y. Basar, M. Itskov, A. Eckstein, Composite laminates: nonlinear interlaminar stress analysis by multi-layer shell elements. Comput. Methods Appl. Mech. Eng. **185**, 367–397 (2000)
102. L. Wang, G. Thierauf, Finite rotations in non-linear analysis of elastic shells. Comput. Struct. **79**, 2357–2367 (2001)
103. J. Pitkäranta, The problem of membrane locking in finite element analysis of cylindrical shells. Numerische Mathematik **61**, 523–542 (1992)
104. Y. Leino, J. Pitkäranta, On the membrane locking of $h - p$ finite elements in a cylindrical shell problem. Int. J. Numer. Methods Eng. **37**, 1053–1070 (1994)
105. K.J. Bathe, A formulation of general shell elements-the use of mixed interpolation of tensorial components. Int. J. Numer. Methods Eng. **22**, 697–722 (1986)
106. F. Brezzi, K.J. Bathe, M. Fortin, Mixed-interpolated elements for Reissner-Mindlin plates. Int. J. Numer. Methods Eng. **28**, 1787–1801 (1989)
107. R.L. Muhanna, New assumed natural strain formulation of the shallow shell element. Commun. Numer. Methods Eng. **9**, 989–1004 (1993)
108. H.C. Huang, E. Hinton, A new nine node degenerated shell element with enhanced membrane and shear interpolation. Int. J. Numer. Methods Eng. **22**, 73–92 (1986)
109. J.C. Simo, T.J.R. Hughes, On the variational foundations of assumed strain methods. J. Appl. Mech. **53**, 51–54 (1986)

110. J.C. Simo, M.S. Rifai, A class of mixed assumed strain methods and the method of incompatible modes. Int. J. Numer. Methods Eng. **29**, 1595–1638 (1990)
111. U. Andelfinger, E. Ramm, EAS-element for two-dimensional, three-dimensional, plate and shell structures and their equivalence to HR-elements. Int. J. Numer. Methods Eng. **36**, 1311–1337 (1993)
112. T.J.R. Hughes, Generalization of selective integration procedures to anisotropic and nonlinear media. Int. J. Numer. Methods Eng. **15**, 1413–1418 (1980)
113. K.C. Park, Improved strain interpolation for curved C^0 elements. Int. J. Numer. Methods Eng. **22**, 281–288 (1986)
114. J.X. Gao, Y.P. Shen, Active control of geometrically nonlinear transient vibration of composite plates with piezoelectric actuators. J. Sound Vib. **264**, 911–928 (2003)
115. I. Kreja, *Geometrically non-linear analysis of layered composite plates and shells*. Habilitation Thesis, Published as Monografie 83, Politechnika Gdańska (2007)
116. J. Pitkäranta, Y. Leino, O. Ovaskainen, J. Piila, Shell deformation states and the finite element method: a benchmark study of cylindrical shells. Comput. Methods Appl. Mech. Eng. **128**, 81–121 (1995)
117. H. Hakula, Y. Leino, J. Pitkäranta, Scale resolution, locking, and high-order finite element modelling of shells. Comput. Methods Appl. Mech. Eng. **133**, 157–182 (1996)
118. J.P. Pontaza, J.N. Reddy, Least-squares finite element formulation for shear-deformable shells. Comput. Methods Appl. Mech. Eng. **194**, 2464–2493 (2005)
119. S. Im, S.N. Atluri, Effects of a piezo-actuator on a finitely deformed beam subjected to general loading. AIAA J. **27**, 1801–1807 (1989)
120. S. Kapuria, P.C. Dumir, Geometrically nonlinear axisymmetric response of thin circular plate under piezoelectric actuation. Commun. Nonlinear Sci. Numer. Simul. **10**, 411–423 (2005)
121. D. Varelis, D.A. Saravanos, Coupled buckling and postbuckling analysis of active laminated piezoelectric composite plates. Int. J. Solids Struct. **41**, 1519–1538 (2004)
122. S. Panda, M.C. Ray, Nonlinear finite element analysis of functionally graded plates integrated with patches of piezoelectric fiber reinforced composite. Finite Elem. Anal. Des. **44**, 493–504 (2008)
123. D. Varelis, D.A. Saravanos, Mechanics and finite element for nonlinear response of active laminated piezoelectric plates. AIAA J. **42**, 1227–1235 (2004)
124. A. Mukherjee, A.S. Chaudhuri, Nonlinear dynamic response of piezolaminated smart beams. Comput. Struct. **83**, 1298–1304 (2005)
125. R. Schmidt, T.D. Vu, Nonlinear dynamic FE simulation of smart piezolaminated structures based on first- and third-order transverse shear deformation theory. Adv. Mater. Res. **79–82**, 1313–1316 (2009)
126. J. Cheng, B. Wang, S.Y. Du, A theoretical analysis of piezoelectric/composite anisotropic laminate with large-amplitude deflection effect, part I: Fundamental equations. Int. J. Solids Struct. **42**, 6166–6180 (2005)
127. H.S. Shen, D.Q. Yang, Nonlinear vibration of functionally graded fiber reinforced composite laminated beams with piezoelectric fiber reinforced composite actuators in thermal environments. Eng. Struct. **90**, 183–192 (2015)
128. V.K. Singh, T.R. Mahapatra, S.K. Panda, Nonlinear flexural analysis of single/doubly curved smart composite shell panels integrated with PFRC actuator. Eur. J. Mech. A/Solids **60**, 300–314 (2016)
129. E. Carrera, An improved Reissner-Mindlin-type model for the electromechanical analysis of multilayered plates including piezo-layers. J. Intell. Mater. Syst. Struct. **8**, 232–248 (1997)
130. S. Kapuria, N. Alam, Zigzag theory for buckling of hybrid piezoelectric beams under electromechanical loads. Int. J. Mech. Sci. **46**, 1–25 (2004)
131. M.C. Ray, J. Shivakumar, Active constrained layer damping of geometrically nonlinear transient vibrations of composite plates using piezoelectric fiber-reinforced composite. Thin-Walled Struct. **47**, 178–189 (2009)
132. S.K. Sarangi, M.C. Ray, Active damping of geometrically nonlinear vibrations of doubly curved laminated composite shells. Compos. Struct. **93**, 3216–3228 (2011)

133. U. Icardi, M.D. Sciuva, Large-deformation and stress analysis of multilayered plates with induced-strain actuators. Smart Mater. Struct. **5**, 140–164 (1996)
134. S. Lentzen, R. Schmidt, A geometrically nonlinear finite element for transient analysis of piezolaminated shells, in *Proceedings Fifth EUROMECH Nonlinear Dynamics Conference* (Eindhoven, Netherlands, 7–12 August 2005), pp. 2492–2500
135. S. Lentzen, P. Klosowski, R. Schmidt, Geometrically nonlinear finite element simulation of smart piezolaminated plates and shells. Smart Mater. Struct. **16**, 2265–2274 (2007)
136. S.Q. Zhang, R. Schmidt, Large rotation theory for static analysis of composite and piezoelectric laminated thin-walled structures. Thin-Walled Struct. **78**, 16–25 (2014)
137. S.Q. Zhang, R. Schmidt, Static and dynamic FE analysis of piezoelectric integrated thin-walled composite structures with large rotations. Compos. Struct. **112**, 345–357 (2014)
138. S.Q. Zhang, R. Schmidt, Large rotation FE transient analysis of piezolaminated thin-walled smart structures. Smart Mater. Struct. **22**, 105025 (2013)
139. J.M.S. Moita, C.M.M. Soares, C.A.M. Soares, Geometrically non-linear analysis of composite structures with integrated piezoelectric sensors and actuators. Compos. Struct. **57**, 253–261 (2002)
140. C.K. Kundu, D.K. Maiti, P.K. Sinha, Post buckling analysis of smart laminated doubly curved shells. Compos. Struct. **81**, 314–322 (2007)
141. P. Dash, B.N. Singh, Nonlinear free vibration of piezoelectric laminated composite plate. Finite Elem. Anal. Des. **45**, 686–694 (2009)
142. M. Amabili, Nonlinear vibrations and stability of laminated shells using a modified first-order shear deformation theory. Eur. J. Mech./ A Solids **68**, 75–87 (2018)
143. M. Amabili, A new third-order shear deformation theory with non-linearities in shear for static and dynamic analysis of laminated doubly curved shells, in *Composite Structures* (2015)
144. F. Alijani, M. Amabili, Non-linear static bending and forced vibrations of rectangular plates retaining non-linearities in rotations and thickness deformation. Int. J. Non-Linear Mech. **67**, 394–404 (2014)
145. F. Alijani, M. Amabili, Effect of thickness deformation on large-amplitude vibrations of functionally graded rectangular plates. Compos. Struct. **113**, 89–107 (2014)
146. M. Amabili, A non-linear higher-order thickness stretching and shear deformation theory for large amplitude vibrations of laminated doubly curved shells. Int. J. Non-Linear Mech. **58**, 57–75 (2014)
147. M. Amabili, J.N. Reddy, A new non-linear higher-order shear deformation theory for large-amplitude vibrations of laminated doubly curved shells. Int. J. Non-Linear Mech. **45**, 409–418 (2010)
148. J. Chróścielewski, P. Klosowski, R. Schmidt, Numerical simulation of geometrically non-linear flexible beam control via piezoelectric layers. ZAMM · Z. Angew. Math. Mech. **77**(Supplement 1), S69–S70 (1997)
149. J. Chróścielewski, P. Klosowski, R. Schmidt, Theory and numerical simulation of nonlinear vibration control of arches with piezoelectric distributed actuators. Mach. Dyn. Probl. **20**, 73–90 (1998)
150. J. Chróścielewski, R. Schmidt, V.A. Eremeyev, Nonlinear finite element modeling of vibration control of plane rod-type structural members with integrated piezoelectric patches, in *Continuum Mechanics and Thermodynamics*, online (2018)
151. S.Q. Zhang, R. Schmidt, Large rotation static and dynamic fe analysis for thin-walled piezolaminated smart structures, in *Proceedings of Shell Structures - Theory and Application (SSTA) 2013*, ed. by W. Pietraszkiewicz, Gdansk, Poland, 16–18 October 2013 (Taylor & Francis Group, London, UK, CRC Press/Balkema, Leiden, The Netherlands, 2013), pp. 477–480
152. M.N. Rao, R. Schmidt, Static and dynamic finite rotation FE-analysis of thin-walled structures with piezoelectric sensor and actuator patches or layers. Smart Mater. Struct. **23**, 095006 (2014)
153. M.N. Rao, R. Schmidt, K.U. Schröder, Geometrically nonlinear static FE-simulation of multilayered magneto-electro-elastic composite structures, in *Composite Structures* (2015)

154. J.S. Yang, Equations for the flexural motion of elastic plates with partially electroded piezoelectric actuators. Smart Mater. Struct. **6**, 485–490 (1997)
155. J.S. Yang, Equations for thick elastic plates with partially electroded piezoelectric actuators and higher order electric fields. Smart Mater. Struct. **8**, 73–82 (1999)
156. S.V. Gopinathan, V.V. Varadan, V.K. Varadan, A review and critique of theories for piezoelectric laminates. Smart Mater. Struct. **9**, 24–48 (2000)
157. V. Cotoni, P. Masson, F. Côté, A finite element for piezoelectric multilayered plates: combined higher-order and piecewise linear c^0 formulation. J. Intell. Mater. Syst. Struct. **17**, 155–166 (2006)
158. M. Kögl, M.A. Bucalem, Analysis of smart laminates using piezoelectric MITC plate and shell elements. Comput. Struct. **83**, 1153–1163 (2005)
159. D. Marinković, H. Köppe, U. Gabbert, Degenerated shell element for geometrically nonlinear analysis of thin-walled piezoelectric active structures. Smart Mater. Struct. **17**, 1–10 (2008)
160. D. Marinković, H. Köppe, U. Gabbert, Aspects of modeling piezoelectric active thin-walled structures. J. Intell. Mater. Syst. Struct. **20**, 1835–1844 (2009)
161. D.F. Nelson, Theory of nonlinear electroacoustics of dielectric, piezoelectric, and pyroelectric crystals. J. Acoust. Soc. Am. **63**(6), 1738–1748 (1978)
162. S.P. Joshi, Non-linear constitutive relations for piezoceramic materials. Smart Mater. Struct. **1**, 80–83 (1992)
163. J.S. Yang, R.C. Batra, A second-order theory for piezoelectric materials. J. Acoust. Soc. Am. **97**(1), 280–288 (1995)
164. M. Kamlah, U. Böhle, Finite element analysis of piezoceramic components taking into account ferroelectric hysteresis behavior. Int. J. Solids Struct. **38**, 605–633 (2001)
165. X. Zhou, A. Chattopadhyay, Nonlinear piezoelectric constitutive relationship and actuation for piezoelectric laminates, in *43rd AIAA/ASME/ASCE/AHS/ASC Structures, Structural Dynamics, and Materials Conference* (Denver, Colorado, 22–25 April 2002), pp. AIAA–2002–1438
166. C.M. Landis, Non-linear constitutive modeling of ferroelectrics. Curr. Opin. Solid State Mater. Sci. **8**, 59–69 (2004)
167. M. Elhadrouz, T.B. Zineb, E. Patoor, Finite element analysis of a multilayer piezoelectric actuator taking into account the ferroelectric and ferroelastic behaviors. Int. J. Eng. Sci. **44**, 996–1006 (2006)
168. L. Ma, Y. Shen, J. Li, H. Zheng, T. Zou, Modeling hysteresis for piezoelectric actuators. J. Intell. Mater. Syst. Struct. **27**(10), 1404–1411 (2016)
169. S. Li, W. Cao, L.E. Cross, The extrinsic nature of nonlinear behavior observed in lead zirconate titanate ferroelectric ceramic. J. Appl. Phys. **69**(10), 7219–7224 (1991)
170. A.J. Masys, W. Ren, G. Yang, B.K. Mukherjee, Piezoelectric strain in lead zirconate titante ceramics as a function of electric field, frequency, and dc bias. J. Appl. Phys. **94**(2), 1155–1162 (2003)
171. S. Klinkel, A phenomenological constitutive model for ferroelastic and ferroelectric hysteresis effects in ferroelectric ceramics. Int. J. Solids Struct. **43**, 7197–7222 (2006)
172. K. Linnemann, S. Klinkel, W. Wagner, A constitutive model for magnetostrictive and piezoelectric materials. Int. J. Solids Struct. **46**, 1149–1166 (2009)
173. P. Tan, L. Tong, A one-dimensional model for non-linear behaviour of piezoelectric composite materials. Compos. Struct. **58**(4), 551–561 (2002)
174. Q.M. Wang, Q. Zhang, B. Xu, R. Liu, L.E. Cross, Nonlinear piezoelectric behavior of ceramic bending mode actuators under strong electric fields. J. Appl. Phys. **86**(6), 3352–3360 (1999)
175. L.Q. Yao, J.G. Zhang, L. Lu, M.O. Lai, Nonlinear dynamic characteristics of piezoelectric bending actuators under strong applied electric. J. Microelectromechanical Syst. **13**(4), 645–652 (2004)
176. D. Sun, L. Tong, D. Wang, An incremental algorithm for static shape control of smart structures with nonlinear piezoelectric actuators. Int. J. Solids Struct. **41**, 2277–2292 (2004)
177. Z.K. Kusculuoglu, T.J. Royston, Nonlinear modeling of composite plates with piezoceramic layers using finite element analysis. J. Sound Vib. **315**, 911–926 (2008)

178. S. Kapuria, M.Y. Yasin, A nonlinear efficient layerwise finite element model for smart piezo-laminated composites under strong applied electric field, in *Smart Materials and Structures* (2013)
179. S. Kapuria, M.Y. Yasin, Active vibration control of piezolaminated composite plates considering strong electric field nonlinearity. AIAA J. **53**(3), 603–616 (2015)
180. M.N. Rao, S. Tarun, R. Schmidt, K.U. Schröder, Finite element modeling and analysis of piezo-integrated composite structures under large applied electric fields. Smart Mater. Struct. **25**, 055044 (2016)
181. L.Q. Yao, J.G. Zhang, L. Lu, M.O. Lai, Nonlinear extension and bending of piezoelectric laminated plate under large applied field actuation. Smart Mater. Struct. **13**, 404–414 (2004)
182. S.Q. Zhang, G.Z. Zhao, S.Y. Zhang, R. Schmidt, X.S. Qin, Geometrically nonlinear FE analysis of piezoelectric laminated composite structures under strong driving electric field. Compos. Struct. **181**, 112–120 (2017)
183. A. Alibeigloo, Free vibration analysis of functionally graded carbon nanotubereinforced composite cylindrical panel embedded in piezoelectric layers by using theory of elasticity. Eur. J. Mech. A/Solids **44**, 104–115 (2014)
184. B.A. Selim, L.W. Zhang, K.M. Liew, Active vibration control of CNT-reinforced composite plates with piezoelectric layers based on Reddy's higher-order shear deformation theory. Compos. Struct. **163**, 350–364 (2017)
185. M. Rafiee, J. Yang, S. Kitipornchai, Large amplitude vibration of carbon nanotube reinforced functionally graded composite beams with piezoelectric layers. Compos. Struct. **96**, 716–725 (2013)
186. R. Ansari, T. Pourashraf, R. Gholami, A. Shahabodini, Analytical solution for nonlinear postbuckling of functionally graded carbon nanotube-reinforced composite shells with piezoelectric layers. Compos. Part B **90**, 267–277 (2016)
187. M.A.R. Loja, C.M.M. Soares, J.I. Barbosa, Analysis of functionally graded sandwich plate structures with piezoelectric skins, using B-spline finite strip method. Compos. Struct. **96**, 606–615 (2013)
188. S. Mikaeeli, B. Behjat, Three-dimensional analysis of thick functionally graded piezoelectric plate using EFG method. Compos. Struct. **154**, 591–599 (2016)
189. Z. Su, G. Jin, T. Ye, Electro-mechanical vibration characteristics of functionally graded piezoelectric plates with general boundary conditions. Int. J. Mech. Sci. **138–139**, 42–53 (2018)
190. M. Derayatifar, M. Tahani, H. Moeenfard, Nonlinear analysis of functionally graded piezoelectric energy harvesters. Compos. Struct. **182**, 199–208 (2017)
191. Y.Q. Wang, Electro-mechanical vibration analysis of functionally graded piezoelectric porous plates in the translation state, in *Acta Astronautica* (2018)
192. M. Krommer, H. Irschik, A Reissner-Mindlin-type plate theory including the direct piezoelectric and the pyroelectric effect. Acta Mechanica **141**, 51–69 (2000)
193. D.P. Zhang, Y.J. Lei, Z.B. Shen, Thermo-electro-mechanical vibration analysis of piezoelectric nanoplates resting on viscoelastic foundation with various boundary conditions. Int. J. Mech. Sci. **131–132**, 1001–1015 (2017)
194. C. Li, J.J. Liu, M. Cheng, X.L. Fan, Nonlocal vibrations and stabilities in parametric resonance of axially moving viscoelastic piezoelectric nanoplate subjected to thermo-electro-mechanical forces. Compos. Part B **116**, 153–169 (2017)
195. M. Arefi, A.M. Zenkour, Thermo-electro-mechanical bending behavior of sandwich nanoplate integrated with piezoelectric face-sheets based on trigonometric plate theory. Compos. Struct. **162**, 108–122 (2017)
196. M. Dehghan, M.Z. Nejad, A. Moosaie, Thermo-electro-elastic analysis of functionally graded piezoelectric shells of revolution: governing equations and solutions for some simple cases. Int. J. Eng. Sci. **104**, 34–61 (2016)
197. A. Alibeigloo, Thermoelastic analysis of functionally graded carbon nanotube reinforced composite cylindrical panel embedded in piezoelectric sensor and actuator layers. Compos. Part B **98**, 225–243 (2016)

198. Z.G. Song, L.W. Zhang, K.M. Liew, Active vibration control of CNT-reinforced composite cylindrical shells via piezoelectric patches. Compos. Struct. **158**, 92–100 (2016)
199. T. Tang, S.D. Felicelli, A multiscale model for electro-thermo-elasto-plastic piezoelectric metal matrix multifunctional composites. Int. J. Eng. Sci. **73**, 1–9 (2013)
200. M. Komijani, J.N. Reddy, M.R. Eslami, Nonlinear analysis of microstructure-dependent functionally graded piezoelectric material actuators. J. Mech. Phys. Solids **63**, 214–227 (2014)
201. M.S. Boroujerdy, M.R. Eslami, Axisymmetric snap-through behavior of Piezo-FGM shallow clamped spherical shells under thermo-electro-mechanical loading. Int. J. Press. Vessel. Pip. **120–121**, 19–26 (2014)
202. M. Vinyas, S.C. Kattimani, Hygrothermal analysis of magneto-electro-elastic plate using 3D finite element analysis. Compos. Struct. **180**, 617–637 (2017)
203. Z.X. Yang, P.F. Dang, Q.K. Han, Z.H. Jin, Natural characteristics analysis of magneto-electro-elastic multilayered plate using analytical and finite element method. Compos. Struct. **185**, 411–420 (2018)
204. S. Razavi, A. Shooshtari, Free vibration analysis of a magneto-electroelastic doubly-curved shell resting on a Pasternak-type elastic foundation. Smart Mater. Struct. **23**, 105003 (9pp), (2014)
205. F. Ebrahimi, A. Jafari, M.R. Barati, Vibration analysis of magneto-electro-elastic heterogeneous porous material plates resting on elastic foundations. Thin-Walled Struct. **119**, 33–46 (2017)
206. H. Ezzin, M. Mkaoir, M.B. Amor, Rayleigh wave behavior in functionally graded magnetoelectro-elastic material. Superlattices Microstruct. **112**, 455–469 (2017)
207. A. Milazzo, Large deflection of magneto-electro-elastic laminated plates. Appl. Math. Model. **38**, 1737–1752 (2014)
208. S.C. Kattimani, M.C. Ray, Smart damping of geometrically nonlinear vibrations of magneto-electro-elastic plates. Compos. Struct. **114**, 51–63 (2014)
209. S. Sahmani, M.M. Aghdam, Nonlocal strain gradient shell model for axial buckling and postbuckling analysis of magneto-electro-elastic composite nanoshells. Compos. Part B **132**, 258–274 (2018)
210. Z. Wang, X. Qin, H.T.Y. Yang, Active suppression of panel flutter with piezoelectric actuators using eigenvector orientation method. J. Sound Vib. **331**, 1469–1482 (2012)
211. Z.G. Song, F.M. Li, Active aeroelastic flutter analysis and vibration control of supersonic composite laminated plate. Compos. Struct. **94**, 702–713 (2012)
212. F.M. Li, Active aeroelastic flutter suppression of a supersonic plate with piezoelectric material. Int. J. Eng. Sci. **51**, 190–203 (2012)
213. M. Mohammadimehr, M. Mehrabi, Electro-thermo-mechanical vibration and stability analyses of double-bonded micro composite sandwich piezoelectric tubes conveying fluid flow. Appl. Math. Model. **60**, 255–272 (2018)
214. Z.G. Song, F.M. Li, E. Carrera, P. Hagedorn, A new method of smart and optimal flutter control for composite laminated panels in supersonic airflow under thermal effects. J. Sound Vib. **414**, 218–232 (2018)
215. D.P. Skinner, R.E. Newnham, L.J.E. Cross, Flexible composite transducers. Mater. Res. Bull. **13**(6), 599–607 (1978)
216. N. Hagood, R. Kindel, K. Ghandi, P. Gaudenzi, Improving transverse actuation of piezoceramics using interdigitated surface electrodes, in *Proceedings of SPIE - Smart Structures and Materials 1993: Smart Structures and Intelligent Systems*, vol. 1917 (Albuquerque, New Mexico, 8 September 1993. SPIE), pp. 341–352
217. A.A. Bent, N.W. Hagood, Piezoelectric fiber composite with integrated electrodes. J. Intell. Mater. Syst. Struct. **8**, 903–919 (1997)
218. W.K. Wilkie, R.G. Bryant, J.W. High, R.L. Fox, R.F. Hellbaum, A. Jalink, B.D. Little, P.H. Mirick, Low-cost piezocomposite actuator for structural control applications, in *SPIE - Smart Structures and Materials 2000: Industrial and Commercial Applications of Smart Structures Technologies*, vol. 3991 (SPIE, 12 June 2000), pp. 323–334

219. R.B. Williams, W.K. Wilkie, D.J. Inman, An overview of composite actuators with piezoceramic fibers, in *Proceedings of IMAC-XX: Conference & Exposition on Structural Dynamics*, vol. 4753 (Los Angeles, CA; United States, 4–7 February 2002), pp. 421–427
220. H.A. Sodano, J. Lloyd, D.J. Inman, An experimental comparison between several active composite actuators for power generation. Smart Mater. Struct. **15**, 1211–1216 (2006)
221. C.R. Bowen, R. Stevens, L.J. Nelson, A.C. Dent, G. Dolman, B. Su, T.W. Button, M.G. Cain, M. Stewart, Manufacture and characterization of high activity piezoelectric fibres. Smart Mater. Struct. **15**, 295–301 (2006)
222. S.C. Choi, J.S. Park, J.H. Kim, Vibration control of pre-twisted rotating composite thin-walled beams with piezoelectric fiber composites. J. Sound Vib. **300**, 176–196 (2007)
223. H.Y. Zhang, Y.P. Shen, Vibration suppression of laminated plates with 1–3 piezoelectric fiber-reinforced composite layers equipped with integrated electrodes. Compos. Struct. **79**, 220–228 (2007)
224. H.A. Sodano, G. Park, D.J. Inman, An investigation into the performance of macro-fiber composites for sensing and structural vibration applications. Mech. Syst. Signal Process. **18**(3), 683–697 (2004)
225. A. Kovalovs, E. Barkanov, S. Gluhihs, Active control of structures using macro-fiber composite (MFC). J. Phys. Conf. Ser. **93**(1) (2007)
226. H.P. Konka, M.A. Wahab, K. Lian, Piezoelectric fiber composite transducers for health monitoring in composite structures. Sens. Actuators A: Phys. **194**, 84–94 (2013)
227. R.B. Williams, D.J. Inman, M.R. Schultz, M.W. Hyer, W.K. Wilkie, Nonlinear tensile and shear behavior of macro fiber composite actuators. J. Compos. Mater. **38**(10), 855–869 (2004)
228. R.B. Williams, *Nonlinear mechanical and actuation characterization of piezoceramic fiber composites*. PhD thesis, Virginia Polytechnic Institute and State University (2004)
229. J.S. Park, J.H. Kim, Analytical development of single crystal macro fiber composite actuators for active twist rotor blades. Smart Mater. Struct. **14**(4), 745 (2005)
230. A. Deraemaeker, S. Benelechi, A. Benjeddou, A. Preumont, Analytical and numerical computation of homogenized properties of MFCs: Application to a composite boom with MFC actuators and sensors, in *Proceedings of the III ECCOMAS Thematic Conference on Smart Structures and Materials* (Gdansk, Poland, 9–11 July 2007)
231. A. Deraemaeker, H. Nasser, A. Benjeddou, A. Preumont, Mixing rules for the piezoelectric properties of macro fiber composites. J. Intell. Mater. Syst. Struct. **20**(12), 1475–1482 (2009)
232. A. Deraemaeker, H. Nasser, Numerical evaluation of the equivalent properties of macro fiber composite (MFC) transducers using periodic homogenization. Int. J. Solids Struct. **47**, 3272–3285 (2010)
233. F. Biscani, H. Nasser, S. Belouettar, E. Carrera, Equivalent electro-elastic properties of macro fiber composite (MFC) transducers using asymptotic expansion approach. Compos. Part B **42**, 444–455 (2011)
234. R.B. Williams, D.J. Inman, W.K. Wilkie, Nonlinear response of the macro fiber composite actuator to monotonically increasing excitation voltage. J. Intell. Mater. Syst. Struct. **17**, 601–608 (2006)
235. J. Schröck, T. Meurer, A. Kugi, Control of a flexible beam actuated by macr-fiber composite patches: II. hysteresis and creep compensation, experimental results. Smart Mater. Struct. **20**, 015016 (2011)
236. Y.X. Li, S.Q. Zhang, R. Schmidt, X.S. Qin, Homogenization for macro-fiber composites using Reissner-Mindlin plate theory, in *Journal of Intelligent Material Systems and Structures* (2016)
237. M.A. Trindade, A. Benjeddou, Finite element characterisation of multilayer d31 piezoelectric macro-fibre composites. Compos. Struct. **151**, 47–57 (2016)
238. M.L. Dano, M. Gakwaya, B. Jullière, Compensation of thermally induced distortion in composite structures using macro-fiber composites. J. Intell. Mater. Syst. Struct. **19**, 225–233 (2008)
239. C.R. Bowen, P.F. Giddings, A.I.T. Salo, H.A. Kim, Modeling and characterization of piezoelectrically actuated bistable composites. IEEE Trans. Ultrason. Ferroelectr. Freq. Control **58**(9), 1737–1750 (2011)

240. L. Ren, A theoretical study on shape control of arbitrary lay-up laminates using piezoelectric actuators. Compos. Struct. **83**, 110–118 (2008)
241. P. Binette, M.L. Dano, G. Gendron, Active shape control of composite structures under thermal loading. Smart Mater. Struct. **18**, 025007 (2009)
242. P.F. Giddings, H.A. Kim, A.I.T. Salo, C.R. Bowen, Modelling of piezoelectrically actuated bistable composites. Mater. Lett. **65**(9), 1261–1263 (2011)
243. O. Bilgen, A. Erturk, D.J. Inman, Analytical and experimental characterization of macro-fiber composite actuated thin clamped-free unimorph benders. J. Vib. Acoust. **132**(5), 051005 (2010)
244. S.Q. Zhang, Y.X. Li, R. Schmidt, Modeling and simulation of macro-fiber composite layered smart structures. Compos. Struct. **126**, 89–100 (2015)
245. S.Q. Zhang, M. Chen, G.Z. Zhao, Z.X. Wang, R. Schmidt, X.S. Qin, Modeling techniques for active shape and vibration control of macro-fiber composite laminated structures. Smart Struct. Syst. **19**(6), 633 (2017)
246. M.S. Azzouz, C. Hall, Nonlinear finite element analysis of a rotating MFC actuator, in *51st AIAA/ASME/ASCE/AHS/ASC Structures, Structural Dynamics, and Materials Conference* (Orlando, Florida, 12–15 April 2010)
247. S.Q. Zhang, Z.X. Wang, X.S. Qin, G.Z. Zhao, R. Schmidt, Geometrically nonlinear analysis of composite laminated structures with multiple macro-fiber composite (MFC) actuators. Compos. Struct. **150**, 62–72 (2016)
248. H.S. Tzou, C.I. Tseng, Distributed modal identification and vibration control of continua: piezoelectric finite element formulation and analysis. J. Dyn. Syst. Meas. Control **113**, 500–505 (1991)
249. H.S. Tzou, C.I. Tseng, Distributed structral indentification and control of shells using districbuted piezoelectric: theory and finite element analysis. Dyn. Control **1**, 297–320 (1991)
250. H.S. Tzou, C.I. Tseng, Distributed vibration control and identification of coupled elastic/piezoelectric systems: finite element formulation and applications. Mech. Syst. Signal Process. **5**, 215–231 (1991)
251. H.S. Tzou, A new distributed sensor and actuator theory for "intelligent" shells. J. Sound Vib. **153**, 335–349 (1992)
252. K.M. Liew, X.Q. He, T.Y. Ng, S. Sivashanker, Active control of FGM plates subjected to a temperature gradient: modeling via finite element method based on FSDT. Int. J. Numer. Methods Eng. **52**, 1253–1271 (2001)
253. Y.K. Kang, H.C. Park, J. Kim, S.B. Choi, Interaction of active and passive vibration control of laminated composite beams with piezoelectric sensors/actuators. Mater. Des. **23**, 277–286 (2002)
254. D. Huang, B. Sun, Approximate analytical solutions of smart composite mindlin beams. J. Sound vib. **244**, 379–394 (2001)
255. B. Sun, D. Huang, Vibration suppression of laminated composite beams with a piezo-electric damping layer. Compos. Struct. **53**, 437–447 (2001)
256. R. Kumar, B.K. Mishra, S.C. Jain, Static and dynamic analysis of smart cylindrical shell. Finite Elem. Anal. Des. **45**, 13–24 (2008)
257. G.G. Sheng, X. Wang, Active control of functionally graded laminated cylindrical shells. Compos. Struct. **90**, 448–457 (2009)
258. S.Q. Zhang, R. Schmidt, Active control for piezoelectric integrated smart structures, in *Proceedings of the 15th International Conference on Experimental Mechanics, Experimental Mechanics: New Trends and Perspectives*, ed. by J.F.S. Gomes, M.A.P. Vaz (Porto, Portugal, 22–27 July 2012), pp. 1029–1030. Edicoes INEGI
259. J.M.S. Moita, C.M.M. Soares, C.A.M. Soares, Active control of forced vibrations in adaptive structures using a higher order model. Compos. Struct. **71**, 349–355 (2005)
260. S.A. Kulkarni, K.M. Bajoria, Finite element modeling of smart plates/shells using higher order shear deformation theory. Compos. Struct. **62**, 41–50 (2003)
261. C.M.A. Vasques, J.D. Rodrigues, Active vibration control of smart piezoelectric beams: comparison of classical and optimal feedback control strategies. Comput. Struct. **84**, 1402–1414 (2006)

262. L. Malgaca, Integration of active vibration control methods with finite element models of smart laminated composite structures. Compos. Struct. **92**, 1651–1663 (2010)
263. S. Raja, P.K. Sinha, G. Prathap, P. Bhattacharya, Influence of one and two dimensional piezoelectric actuation on active vibration control of smart panels. Aerosp. Sci. Technol. **209–216**, 6 (2002)
264. S. Valliappan, K. Qi, Finite element analysis of a 'smart' damper for seismic structural control. Comput. Struct. **81**, 1009–1017 (2003)
265. S.X. Xu, T.S. Koko, Finite element analysis and design of actively controlled piezoelectric smart structures. Finite Elem. Anal. Des. **40**, 241–262 (2004)
266. G.E. Stavroulakis, G. Foutsitzi, E. Hadjigeorgiou, D. Marinova, C.C. Baniotopoulos, Design and robust optimal control of smart beams with application on vibrations suppression. Adv. Eng. Softw. **36**, 806–813 (2005)
267. X.J. Dong, G. Meng, J.C. Peng, Vibration control of piezoelectric smart structures based on system identification technique: numerical simulation and experimental study. J. Sound Vib. **297**, 680–693 (2006)
268. T. Roy, D. Chakraborty, Optimal vibration control of smart fiber reinforced composite shell structures using improved genetic algorithm. J. Sound Vib. **319**, 15–40 (2009)
269. C.Q. Chen, Y.P. Shen, Optimal control of active structures with piezoelectric modal sensors and actuators. Smart Mater. Struct. **6**, 403–409 (1997)
270. J.C. Lin, M.H. Nien, Adaptive control of a composite cantilever beam with piezoelectric damping-modal actuators/sensors. Compos. Struct. **70**, 170–176 (2005)
271. P. Bhattacharya, H. Suhail, P.K. Sinha, Finite element analysis and distributed control of laminated composite shells using LQR/IMSC approach. Aerosp. Sci. Technol. **6**, 273–281 (2002)
272. T.C. Manjunath, B. Bandyopadhyay, Vibration control of Timoshenko smart structures using multirate output feedback based discrete sliding mode control for SISO systems. J. Sound Vib. **326**, 50–74 (2009)
273. S.Q. Zhang, H.N. Li, R. Schmidt, Unknown disturbance estimation and compensation using PI observer for active control of smart beams, in *Proceedings of the XI International Conference on Recent Advances in Structural Dynamics*, ed. by E. Rustighi, et al. (Pisa, Italy, 1–3 July 2013)
274. S.Q. Zhang, H.N. Li, R. Schmidt, P.C. Müller, Disturbance rejection control for vibration suppression of piezoelectric laminated thin-walled structures. J. Sound Vib. **333**, 1209–1223 (2014)
275. S.Q. Zhang, X.Y. Zhang, H.L. Ji, S.S. Ying, R. Schmidt, A refined disturbance rejection control for vibration suppression of smart structures under unknown disturbances, in *Journal of Low Frequency Noise, Vibration and Active Control* (2019)
276. X.Y. Zhang, R.X. Wang, S.Q. Zhang, Z.X. Wang, X.S. Qin, R. Schmidt, Generalized-disturbance rejection control for vibration suppression of piezoelectric laminated flexible structures. Shock Vib. **2018**, ID 1538936 (2018)
277. Y.H. Zhou, J. Wang, Vibration control of piezoelectric beam-type plates with geometrically nonlinear deformation. Int. J. Non-Linear Mech. **39**, 909–920 (2004)
278. D.T. Vu, *Geometrically nonlinear higher-oder shear deformation FE analysis of thin-walled smart structures*. Ph.D. Thesis, RWTH Aachen University (2011)
279. G.S. Lee, System identification and control of smart structures using neural networks. Acra Astronautica **38**, 269–276 (1996)
280. T. Han, L. Acar, A neural network based approach for the identification and optimal control of a cantilever plate, in *Proceedings of the American Control Conference, Albuquerque, New Mexico* (Albuquerque, New Mexico, June 1997), pp. 232–236
281. M.T. Valoor, K. Chandrashekhara, S. Agarwal, Self-adaptive vibration control of smart composite beams using recurrent neural architecture. Int. J. Solids Struct. **38**, 7857–7874 (2001)
282. S.H. Youn, J.H. Han, I. Lee, Neuro-adaptive vibration control of composite beams subject to sudden delamination. J. Sound Vib. **238**, 215–231 (2000)

283. R. Kumar, S.P. Singh, H.N. Chandrawat, MIMO adaptive vibration control of smart structures with quickly varying parameters: neural networks vs classical control approach. J. Sound Vib. **307**, 639–661 (2007)
284. Z. Qiu, X. Zhang, C. Ye, Vibration suppression of a flexible piezoelectric beam using BP neural network control. Acta Mechanica Solida Sinica **25**, 417–428 (2012)
285. R. Jha, C. He, Neural and converntional adaptive predictive controllers for smart structures, in *44th AIAA/ASME/ASCE/AHS Structures, Structural Dynamics, and Materials Conference*, number AIAA 2003-1808, Norfolk, Virginia, 7–10 April 2003. American Institute of Aeronautics and Astronautics, Inc
286. A.H.N. Shirazi, H.R. Owji, M. Rafeeyan, Active vibration control of an FGM rectangular plate using fuzzy logic controllers. Proc. Eng. **14**, 3019–3026 (2011)
287. G.L. Abreu, J.F. Ribeiro, A self-organizing fuzzy logic controller for the active control of flexible structures using piezoelectric actuators. Appl. Soft Comput. **1**, 271–283 (2002)

Chapter 3
Geometrically Nonlinear Theories

Abstract This chapter starts with discussing various hypotheses, and the differences between these hypotheses are outlined. Afterwards, the mathematical preliminaries, including position vectors, covariant and contravariant base vectors, Christoffel symbols, shifter tensor, curvature tensor, etc., will be defined and discussed. Based on the FOSD hypothesis, through-thickness displacement distribution is assumed, where six parameters are introduced. Using these predefined quantities, Green-Lagrange strain tensor with fully geometrically nonlinear strain-displacement relations is developed in terms of six parameters for geometrically nonlinear theory with unrestricted finite rotations (LRT56). Imposing different assumptions, various simplified nonlinear strain-displacement relations are developed for the theories of von Kármán type nonlinear (RVK5), moderate rotation nonlinear (MRT5), fully geometrically nonlincar with moderate rotations (LRT5).

3.1 Shear Deformation Hypotheses

The FE method with 3-D solid elements is one of the possible solutions for modeling of thin-walled composite and smart structures. Even though the thickness of plates and shells are very small compared to the in-plane dimensions, the elements through the thickness direction must reach a certain number to ensure the computation accuracy. Therefore, using 3-D solid element for modeling of thin-walled smart structures certainly results in large model size and high computation time. Because of small thickness in thin-walled plate and shell structures, FE methods with 2-D surface elements based on various hypotheses (shown in Fig. 2.1) are more frequently used in numerical analysis. The main advantage of 2-D FE models is that less computation time is needed due to small size of the models compared to 3-D ones, but they are still retaining a relatively high accuracy. For beam structures, 2-D surface element

S.-Q. Zhang, *Nonlinear Analysis of Thin-Walled Smart Structures*, Springer Tracts in Mechanical Engineering, https://doi.org/10.1007/978-981-15-9857-9_3

can further be simplified to 1-D beam element using Euler-Bernoulli, Timoshenko or higher-order beam hypothesis.

The simplest hypothesis for plates and shells is the Kirchhoff-Love theory, which is also known as classical lamination theory (CLT), see the Refs. [1–7]. The classical lamination theory assumes that the straight lines normal to the mid-surface in the undeformed configuration remain straight and normal after deformation. The Euler-Bernoulli beam theory has a same assumption as that in classical theory. Due to the assumption imposed by classical theory, there is no transverse shear strain in the mathematical model, which may result in an inadequate prediction of the elastic behavior of layered composite and smart structures.

In order to introduce transverse shear strain, Reissner-Mindlin plate/shell theory [8] known as FOSD hypothesis was proposed. The FOSD hypothesis assumes that the straight lines normal to the mid-surface in the undeformed configuration remain straight after deformation, but not necessarily normal. By the assumption, there exists an additional angle between the FOSD line and CLT line, defined as transverse shear strain. Analogously, the FOSD plate/shell theory can be treated as an extension of the Timoshenko beam theory. It can seen that both the classical theory and Reissner-Mindlin theory assume a linear variation of displacement through the shell thickness.

For moderately thick structures, classical theory and the FOSD theory may not be accurate enough. In order to deal with thick structures, SOSD (Second-order Shear Deformation), TOSD or other HOSD hypotheses were proposed. Due to different higher-order hypotheses, the through-thickness displacements can be distributed quadratically, cubicly or other higher-order functions, see Fig. 2.1. With these higher-order shear deformation hypotheses, one can satisfy zero transverse shear strains at outer surfaces while the transverse shear strains vary nonlinearly inside the structure, which is a more practical way that structures usually occur.

Concerning laminated shell structures made of different materials, all the above mentioned hypotheses exist inter-layer transverse shear stress discontinuity. *Zigzag* theory assumes independent transverse shear strain for each layer. This satisfies the inter-layer shear stress continuity. The *first-order zigzag* theory describes the through-thickness displacement as a fold line. For more accurately, *second-* or *third-order zigzag* hypothesis can be employed.

3.2 Mathematical Preliminaries

3.2.1 Introduction of Coordinates

Two coordinate systems are introduced for the mathematical modeling, as shown in Fig. 3.1. One is the Cartesian coordinate system, represented by X^1, X^2 and X^3, acting as global coordinate system. The other one is the curvilinear coordinate system represented by Θ^1, Θ^2 and Θ^3, acting as convective coordinate system. The global

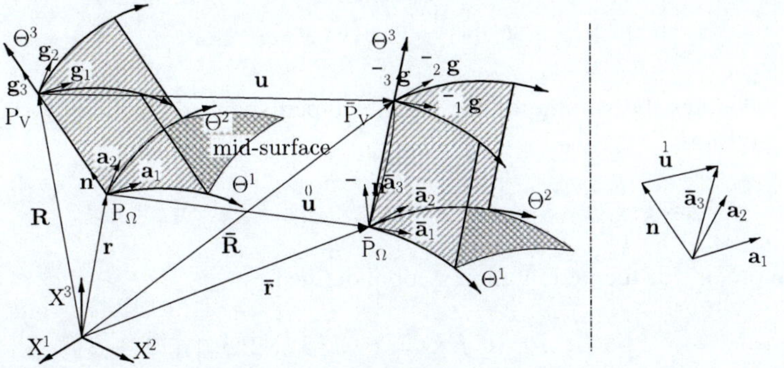

Fig. 3.1 Definition of base vectors

coordinate system is usually fixed, while the convective coordinate system is set on structures. The convective coordinate system can be plate, cylindrical, spherical or any other coordinates. The position vector of an arbitrary point (P_V) in the shell space is denoted by $\boldsymbol{R}(\Theta^1, \Theta^2, \Theta^3)$, while $\boldsymbol{r}(\Theta^1, \Theta^2)$ refers to that of an arbitrary point (P_Ω) at the mid-surface.

In order to present the structural deformation, two configurations are defined, namely the *undeformed configuration* and the *deformed configuration*, as shown in Fig. 3.1. The undeformed configuration is shown in the left part of the figure, while the deformed configuration is shown in the middle part of the figure. Furthermore, the right hand side of the figure shows the rotation of the Θ^3-line. An arbitrary point in the shell space and at the mid-surface is denoted by P_V and P_Ω, respectively. In this report, the Latin indices vary from 1 to 3, whereas the Greek indices only take 1 or 2.

3.2.2 *Base Vectors and Metric Tensor in Shell Space*

Considering an arbitrary point P_V in the undeformed shell space, the covariant base vectors $\boldsymbol{g}_i$ are defined as the tangent of the coordinate lines, expressed by

$$\boldsymbol{g}_i = \frac{\partial \boldsymbol{R}}{\partial \Theta^i} = \boldsymbol{R}_{,i}\,, \tag{3.1}$$

where the subscript “, i” represents the spatial derivative with respect to Θ^i. Because the coordinate lines can be arbitrarily defined, the base vectors $\boldsymbol{g}_i$ may not be perpendicular with each other, like the Cartesian coordinate system. To avoid complex computation problems, we introduce contravariant base vectors $\boldsymbol{g}^i$, which are determined by means of the vector products of covariant base vectors

$$\boldsymbol{g}^i = \frac{1}{V} e^{ijk} \boldsymbol{g}_j \times \boldsymbol{g}_k \,, \tag{3.2}$$

where V denotes the volume of the parallelepiped spanned by the covariant base vectors, given as

$$V = \boldsymbol{g}_1 \cdot (\boldsymbol{g}_2 \times \boldsymbol{g}_3) = \boldsymbol{g}_2 \cdot (\boldsymbol{g}_3 \times \boldsymbol{g}_1) = \boldsymbol{g}_3 \cdot (\boldsymbol{g}_1 \times \boldsymbol{g}_2) \,. \tag{3.3}$$

Furthermore, e^{ijk} is the permutation symbol defined as

$$e^{ijk} = \begin{cases} 1 & \text{for } (i, j, k) = (1, 2, 3), (2, 3, 1), (3, 1, 2) \\ -1 & \text{for } (i, j, k) = (1, 3, 2), (3, 2, 1), (2, 1, 3) \\ 0 & \text{others} \end{cases} \tag{3.4}$$

The scalar product of the covariant and contravariant base vectors results in respectively covariant and contravariant metric tensors as

$$g_{ij} = g_{ji} = \boldsymbol{g}_i \cdot \boldsymbol{g}_j \,, \tag{3.5}$$

$$g^{ij} = g^{ji} = \boldsymbol{g}^i \cdot \boldsymbol{g}^j \,. \tag{3.6}$$

The mixed scalar product of the covariant and contravariant base vectors yields

$$\boldsymbol{g}_i \cdot \boldsymbol{g}^j = \delta_i^j \,, \tag{3.7}$$

in which δ_i^j represent the Kronecker delta, given as

$$\delta_i^j = \begin{cases} 1 & \text{for } i = j \\ 0 & \text{for } i \neq j \end{cases} \,. \tag{3.8}$$

The derivatives of the covariant and contravariant base vectors are

$$\boldsymbol{g}_{i,j} = \Gamma_{ijk} \boldsymbol{g}^k = \Gamma_{ij}^k \boldsymbol{g}_k \,, \tag{3.9}$$

$$\boldsymbol{g}^k{}_{,j} = -\Gamma_{ij}^k \boldsymbol{g}^i \,, \tag{3.10}$$

where Γ_{ijk} and Γ_{ij}^k represent respectively the Christoffel symbols of the first and second kind. The computations of the Christoffel symbols are

$$\Gamma_{ij}^k = \Gamma_{ji}^k = \boldsymbol{g}_{i,j} \cdot \boldsymbol{g}^k = -\boldsymbol{g}_i \cdot \boldsymbol{g}^k{}_{,j} \,, \tag{3.11}$$

$$\Gamma_{ijk} = \Gamma_{jik} = \boldsymbol{g}_{i,j} \cdot \boldsymbol{g}_k \,. \tag{3.12}$$

3.2.3 Base Vectors and Metric Tensor at Mid-surface

In plate and shell structures, a reference surface is assumed to represent the solid structure. The reference surface is where the smallest in-plane deformation energy occurs compared with surfaces through the thickness direction. Sometimes, the reference surface is named as mid-surface, which is not necessarily in the middle position. The base vectors for point P_Ω at the mid-surface in the undeformed configuration are given by

$$\boldsymbol{a}_\alpha = \frac{\partial \boldsymbol{r}}{\partial \Theta^\alpha} = \boldsymbol{r}_{,\alpha} , \tag{3.13}$$

$$\boldsymbol{a}_3 = \boldsymbol{n} = \frac{\boldsymbol{a}_1 \times \boldsymbol{a}_2}{\|\boldsymbol{a}_1 \times \boldsymbol{a}_2\|} , \tag{3.14}$$

where $\| \cdot \|$ represent the Euclidean norm. From the definition, it can be clearly seen that the base vector $\boldsymbol{a}_3$ in the thickness direction is a unit vector and normal to the plane formed by $(\boldsymbol{a}_1, \boldsymbol{a}_2)$. The contravariant base vectors for point P_Ω at the mid-surface in the undeformed configuration are similarly obtained as

$$\boldsymbol{a}^i = \frac{1}{V} e^{ijk} \boldsymbol{a}_j \times \boldsymbol{a}_k , \tag{3.15}$$

The scalar product of the covariant and contravariant base vectors at the reference surface will be

$$\boldsymbol{a}_\alpha \cdot \boldsymbol{a}_\beta = a_{\alpha\beta} , \tag{3.16}$$

$$\boldsymbol{a}^\alpha \cdot \boldsymbol{a}^\beta = a^{\alpha\beta} . \tag{3.17}$$

Here $a_{\alpha\beta}$ and $a^{\alpha\beta}$ respectively represent the covariant and contravariant metric tensors at the mid-surface. Analogously, the mixed scalar product of covariant and contravariant base vectors at the mid-surface can be obtained as

$$\boldsymbol{a}_i \cdot \boldsymbol{a}^j = \delta_i^j . \tag{3.18}$$

From the definition of the vector $\boldsymbol{n}$, we know that $\boldsymbol{n}$ is a unit vector and perpendicular to the plane formed by $(\boldsymbol{a}_1, \boldsymbol{a}_2)$. Therefore, we can get the relations as

$$\boldsymbol{a}_\alpha \cdot \boldsymbol{n} = 0 , \tag{3.19}$$

$$\boldsymbol{n} \cdot \boldsymbol{n} = 1 . \tag{3.20}$$

Taking the derivative of Eqs. (3.19) and (3.20) with respect to Θ^β one obtains

$$\boldsymbol{a}_{\alpha,\beta} \cdot \boldsymbol{n} + \boldsymbol{a}_\alpha \cdot \boldsymbol{n}_{,\beta} = 0 , \tag{3.21}$$

$$\boldsymbol{n} \cdot \boldsymbol{n}_{,\beta} = 0 . \tag{3.22}$$

The derivative of the covariant and contravariant base vectors of point P_Ω, $\boldsymbol{a}_\alpha$ and $\boldsymbol{n}$, with respect to Θ^β can be obtained as

$$\boldsymbol{a}_{\alpha,\beta} = \Gamma^{\delta}_{\alpha\beta}\boldsymbol{a}_\delta + b_{\alpha\beta}\boldsymbol{n}\,, \tag{3.23}$$

$$\boldsymbol{a}^{\alpha}{}_{,\beta} = -\Gamma^{\alpha}_{\delta\beta}\boldsymbol{a}^{\delta} + b^{\alpha}_{\beta}\boldsymbol{n}\,, \tag{3.24}$$

$$\boldsymbol{n}_{,\beta} = -b^{\delta}_{\beta}\boldsymbol{a}_\delta = -b_{\lambda\beta}\boldsymbol{a}^{\lambda}\,. \tag{3.25}$$

Here, $b_{\alpha\beta}$ and b^α_β are the covariant and mixed components of the curvature tensor, respectively, which can be calculated by

$$b_{\alpha\beta} = \boldsymbol{a}_{\alpha,\beta}\cdot\boldsymbol{n} = -\boldsymbol{a}_\alpha\cdot\boldsymbol{n}_{,\beta}\,, \tag{3.26}$$

$$b^{\alpha}_{\beta} = \boldsymbol{a}^{\alpha}{}_{,\beta}\cdot\boldsymbol{n} = -\boldsymbol{a}^{\alpha}\cdot\boldsymbol{n}_{,\beta}\,. \tag{3.27}$$

The relations between the covariant and mixed components of the curvature tensor can be obtained as

$$b^{\lambda}_{\alpha} = a^{\beta\lambda}b_{\alpha\beta}\,. \tag{3.28}$$

3.2.4 Quantities in Deformed Configurations

From Fig. 3.1, two configurations are defined in mathematical theory description, i.e. deformed and undeformed configuration. The quantities introduced in the above subsections are in the undeformed configuration. Using the same notations, but with an overbar, are used for the base vectors and geometric quantities in the deformed configuration, which is shown in the middle part of Fig. 3.1. Thus, the base vectors in the undeformed and deformed configurations are defined and listed in Table 3.1.

Table 3.1 Base vectors in the undeformed and deformed configurations

Name	Undeformed	Deformed
Position vector in the shell space	$\boldsymbol{R}$	$\bar{\boldsymbol{R}}$
Position vector at the mid-surface	$\boldsymbol{r}$	$\bar{\boldsymbol{r}}$
Covariant base vectors in the shell space	$\boldsymbol{g}_1, \boldsymbol{g}_2, \boldsymbol{g}_3$	$\bar{\boldsymbol{g}}_1, \bar{\boldsymbol{g}}_2, \bar{\boldsymbol{g}}_3$
Covariant base vectors at the mid-surface	$\boldsymbol{a}_1, \boldsymbol{a}_2, \boldsymbol{a}_3(\boldsymbol{n})$	$\bar{\boldsymbol{a}}_1, \bar{\boldsymbol{a}}_2, \bar{\boldsymbol{a}}_3$
Contravariant base vectors in the shell space	$\boldsymbol{g}^1, \boldsymbol{g}^2, \boldsymbol{g}^3$	$\bar{\boldsymbol{g}}^1, \bar{\boldsymbol{g}}^2, \bar{\boldsymbol{g}}^3$
Contravariant base vectors at the mid-surface	$\boldsymbol{a}^1, \boldsymbol{a}^2, \boldsymbol{a}^3(\boldsymbol{n})$	$\bar{\boldsymbol{a}}^1, \bar{\boldsymbol{a}}^2, \bar{\boldsymbol{a}}^3$

3.3 Kinematics of Shell Structures

3.3.1 Through-Thickness Displacement Distribution

According to the geometric relations in Fig. 3.1, the position vector of point $\mathrm{P_V}$ in the undeformed configuration can be expressed by the base vectors and position vector at the mid-surface as

$$\boldsymbol{R} = \boldsymbol{r} + \Theta^3 \boldsymbol{n} \,. \tag{3.29}$$

Due to the FOSD hypothesis that straight lines along the thickness direction remain straight, but not necessarily normal to the mid-surface of deformed configuration, the position vector of $\bar{\mathrm{P}}_\mathrm{V}$ in the deformed configuration is

$$\bar{\boldsymbol{R}} = \bar{\boldsymbol{r}} + \Theta^3 \bar{\boldsymbol{a}}_3 \,. \tag{3.30}$$

Furthermore, because of the geometric relation of the FOSD hypothesis, the displacement vector $\boldsymbol{u}$ is defined as

$$\boldsymbol{u} = \bar{\boldsymbol{R}} - \boldsymbol{R} = \overset{0}{\boldsymbol{u}} + \Theta^3 \overset{1}{\boldsymbol{u}} \,. \tag{3.31}$$

Equation (3.31) shows that the displacement is linearly distributed through the thickness direction. Here, $\overset{0}{\boldsymbol{u}}$ denotes the translational displacement vector at the mid-surface, and $\overset{1}{\boldsymbol{u}}$ is the rotational displacement vector, which describes the rotation of the unit normal vector from $\boldsymbol{n}$ to $\bar{\boldsymbol{a}}_3$. They are respectively obtained as

$$\overset{0}{\boldsymbol{u}} = \bar{\boldsymbol{r}} - \boldsymbol{r} \,, \tag{3.32}$$

$$\overset{1}{\boldsymbol{u}} = \bar{\boldsymbol{a}}_3 - \boldsymbol{n} \,. \tag{3.33}$$

Taking the derivative of Eqs. (3.32) and (3.33) with respect to Θ^α yields

$$\overset{0}{\boldsymbol{u}}_{,\alpha} = \bar{\boldsymbol{a}}_\alpha - \boldsymbol{a}_\alpha \,, \tag{3.34}$$

$$\overset{1}{\boldsymbol{u}}_{,\alpha} = \bar{\boldsymbol{a}}_{3,\alpha} - \boldsymbol{n}_{,\alpha} \,. \tag{3.35}$$

Further, the covariant and contravariant components of the translational displacement vector $\overset{0}{\boldsymbol{u}}$ and the rotational displacement vector $\overset{1}{\boldsymbol{u}}$ can be defined as

$$\overset{0}{\boldsymbol{u}} = \overset{0}{v}_\alpha \boldsymbol{a}^\alpha + \overset{0}{v}_3 \boldsymbol{n} = \overset{0}{v}{}^\alpha \boldsymbol{a}_\alpha + \overset{0}{v}{}^3 \boldsymbol{n} \,, \tag{3.36}$$

$$\overset{1}{\boldsymbol{u}} = \overset{1}{v}_\alpha \boldsymbol{a}^\alpha + \overset{1}{v}_3 \boldsymbol{n} = \overset{1}{v}{}^\alpha \boldsymbol{a}_\alpha + \overset{1}{v}{}^3 \boldsymbol{n} \,. \tag{3.37}$$

Here, the six covariant components are considered as six independent kinematic parameters, among which the first three parameters, $\overset{0}{v}_1$, $\overset{0}{v}_2$, $\overset{0}{v}_3$, are the translational displacements at the mid-surface, and the last three parameters, $\overset{1}{v}_1$, $\overset{1}{v}_2$, $\overset{1}{v}_3$, are the generalized rotational displacements, i.e. the projections of $\overset{1}{\boldsymbol{u}}$ in the contravariant base vector triad of the undeformed configuration. The sixth parameter $\overset{1}{v}_3$ is usually neglected in the linear or simplified nonlinear shell theories, due to the assumption of small or moderate rotations occurring in structures. However, when structures undergo large displacements and rotations, $\overset{1}{v}_3$ is no longer small. Therefore the sixth parameter $\overset{1}{v}_3$ must be considered in large rotation theory.

Using the covariant components of the vectors $\overset{0}{\boldsymbol{u}}$ and $\overset{1}{\boldsymbol{u}}$, Eq. (3.31) can also be re-written in scalar form as

$$v_\alpha(\Theta^1, \Theta^2, \Theta^3) = \overset{0}{v}_\alpha(\Theta^1, \Theta^2) + \Theta^3 \overset{1}{v}_\alpha(\Theta^1, \Theta^2)\,, \tag{3.38}$$

$$v_3(\Theta^1, \Theta^2, \Theta^3) = \overset{0}{v}_3(\Theta^1, \Theta^2) + \Theta^3 \overset{1}{v}_3(\Theta^1, \Theta^2)\,. \tag{3.39}$$

Considering the covariant components in Eqs. (3.36) and (3.37), the derivative with respect to Θ^β are

$$\begin{aligned} \overset{n}{\boldsymbol{u}}_{,\beta} &= \overset{n}{v}_{\alpha,\beta}\boldsymbol{a}^\alpha + \overset{n}{v}_\alpha \boldsymbol{a}^\alpha{}_{,\beta} + \overset{n}{v}_{3,\beta}\boldsymbol{n} + \overset{n}{v}_3 \boldsymbol{n}_{,\beta} \\ &= \left(\overset{n}{v}_{\lambda,\beta} - \Gamma^\alpha_{\lambda\beta}\overset{n}{v}_\alpha - b_{\lambda\beta}\overset{n}{v}_3\right)\boldsymbol{a}^\lambda + \left(\overset{n}{v}_{3,\beta} + b^\alpha_\beta \overset{n}{v}_\alpha\right)\boldsymbol{n}\,. \end{aligned} \tag{3.40}$$

Alternatively, the derivatives of (3.36) and (3.37) with respect to Θ^β using the contravariant components are obtained as

$$\begin{aligned} \overset{n}{\boldsymbol{u}}_{,\beta} &= \overset{n}{v}{}^\alpha{}_{,\beta}\boldsymbol{a}_\alpha + \overset{n}{v}{}^\alpha \boldsymbol{a}_{\alpha,\beta} + \overset{n}{v}{}^3{}_{,\beta}\boldsymbol{n} + \overset{n}{v}{}^3 \boldsymbol{n}_{,\beta} \\ &= \left(\overset{n}{v}{}^\lambda{}_{,\beta} + \Gamma^\lambda_{\alpha\beta}\overset{n}{v}{}^\alpha - b^\lambda_\beta \overset{n}{v}{}^3\right)\boldsymbol{a}_\lambda + \left(\overset{n}{v}{}^3{}_{,\beta} + b_{\alpha\beta}\overset{n}{v}{}^\alpha\right)\boldsymbol{n}\,. \end{aligned} \tag{3.41}$$

We introduce the covariant and contravariant derivatives, represented by the subscript "|". The covariant and contravariant derivatives with to Θ^β are defined as

$$\overset{n}{v}_{\lambda|\beta} = \overset{n}{v}_{\lambda,\beta} - \Gamma^\alpha_{\lambda\beta}\overset{n}{v}_\alpha\,, \tag{3.42}$$

$$\overset{n}{v}{}^\lambda{}_{|\beta} = \overset{n}{v}{}^\lambda{}_{,\beta} + \Gamma^\lambda_{\alpha\beta}\overset{n}{v}{}^\alpha\,. \tag{3.43}$$

Further, introducing the following abbreviations

$$\overset{n}{\varphi}_{\lambda\beta} = \overset{n}{v}_{\lambda|\beta} - b_{\lambda\beta}\overset{n}{v}_3 , \tag{3.44}$$

$$\overset{n}{\varphi}{}^{\lambda}_{\beta} = \overset{n}{v}{}^{\lambda}{}_{|\beta} - b^{\lambda}_{\beta}\overset{n}{v}{}^3 , \tag{3.45}$$

$$\overset{n}{\varphi}_{3\beta} = \overset{n}{v}_{3,\beta} + b^{\alpha}_{\beta}\overset{n}{v}_{\alpha} , \tag{3.46}$$

$$\overset{n}{\varphi}{}^{3}_{\beta} = \overset{n}{v}{}^3{}_{,\beta} + b_{\alpha\beta}\overset{n}{v}{}^{\alpha} , \tag{3.47}$$

Equation (3.40) can be re-written as

$$\begin{aligned} \overset{n}{\boldsymbol{u}}_{,\beta} &= \overset{n}{\varphi}_{\lambda\beta}\boldsymbol{a}^{\lambda} + \overset{n}{\varphi}_{3\beta}\boldsymbol{n} \\ \text{or} \quad \overset{n}{\boldsymbol{u}}_{,\beta} &= \overset{n}{\varphi}{}^{\lambda}_{\beta}\boldsymbol{a}_{\lambda} + \overset{n}{\varphi}{}^{3}_{\beta}\boldsymbol{n} . \end{aligned} \tag{3.48}$$

Here, the overhead letter n assumes only the value 0 or 1. From Eqs. (3.36) and (3.37), it can be concluded that $\overset{n}{v}_3 = \overset{n}{v}{}^3$. Therefore, the relations between the abbreviations above can be obtained as

$$\overset{n}{\varphi}{}^{\lambda}_{\beta} = a^{\lambda\alpha}\overset{n}{\varphi}_{\alpha\beta} , \tag{3.49}$$

$$\overset{n}{\varphi}{}^{3}_{\beta} = a^{33}\overset{n}{\varphi}_{3\beta} = \overset{n}{\varphi}_{3\beta} . \tag{3.50}$$

3.3.2 *Shifter Tensor*

The shifter tensor represents the coefficients generated due to the transformation from three-dimensional space to two-dimensional space, which is defined by the tensor product of base vectors at the mid-surface and in the shell space as

$$\boldsymbol{\mu} = \boldsymbol{g}_i \otimes \boldsymbol{a}^i = \mu^j_i \boldsymbol{a}_j \otimes \boldsymbol{a}^i = \mu^{\delta}_{\lambda}\boldsymbol{a}_{\delta} \otimes \boldsymbol{a}^{\lambda} + \boldsymbol{a}_3 \otimes \boldsymbol{a}_3 , \tag{3.51}$$

$$\boldsymbol{\mu}^{\mathrm{T}} = \boldsymbol{a}^i \otimes \boldsymbol{g}_i = \mu^j_i \boldsymbol{a}^i \otimes \boldsymbol{g}_j = \mu^{\delta}_{\lambda}\boldsymbol{a}^{\lambda} \otimes \boldsymbol{a}_{\delta} + \boldsymbol{a}_3 \otimes \boldsymbol{a}_3 . \tag{3.52}$$

Here $\otimes$ represents the tensor product, μ^j_i denote the components of the shifter tensor $\boldsymbol{\mu}$.

The components of the shifter tensor are obtained by taking the spatial derivative of position vector given in (3.29) with respect to Θ^i and using (3.25)

$$\begin{aligned} \boldsymbol{g}_{\alpha} &= \boldsymbol{a}_{\alpha} + \Theta^3 \boldsymbol{n}_{,\alpha} = \left(\delta^{\delta}_{\alpha} - b^{\delta}_{\alpha}\Theta^3\right)\boldsymbol{a}_{\delta} = \mu^{\delta}_{\alpha}\boldsymbol{a}_{\delta} , \\ \boldsymbol{g}_3 &= \boldsymbol{a}_3 = \mu^3_3 \boldsymbol{a}_3 , \end{aligned} \tag{3.53}$$

Therefore, the components of the shifter tensor are expressed as

$$\mu^j_i = \begin{bmatrix} 1 - \Theta^3 b^1_1 & -\Theta^3 b^2_1 & 0 \\ -\Theta^3 b^1_2 & 1 - \Theta^3 b^2_2 & 0 \\ 0 & 0 & 1 \end{bmatrix} . \tag{3.54}$$

We further define the determinant of the shifter tensor, μ, as

$$\begin{aligned}\mu &= \det[\mu_i^j] = 1 - \Theta^3\left(b_1^1 + b_2^2\right) + \left(\Theta^3\right)^2\left(b_1^1 b_2^2 - b_1^2 b_2^1\right) \\ &= 1 - 2H\Theta^3 + K\left(\Theta^3\right)^2 ,\end{aligned} \tag{3.55}$$

where H and K denote respectively the mean and Gaussian curvature of the surface.

Using the shifter tensor, the volume element

$$\mathrm{dV} = (\boldsymbol{g}_1 \times \boldsymbol{g}_2) \cdot \boldsymbol{g}_3\, \mathrm{d}^1\mathrm{d}^2\mathrm{d}^3 = \sqrt{g}\, \mathrm{d}^1\mathrm{d}^2\mathrm{d}^3 \tag{3.56}$$

can be related to the surface element as

$$\mathrm{dV} = \mu\, \mathrm{d}^3\, \mathrm{d}\Omega \tag{3.57}$$

where the surface area element is given by

$$\mathrm{d}\Omega = |\boldsymbol{a}_1 \times \boldsymbol{a}_2|\, \mathrm{d}^1\mathrm{d}^2 = \sqrt{a}\, \mathrm{d}^1\mathrm{d}^2 , \tag{3.58}$$

in which

$$g = \det[g_{ij}] , \quad a = \det[a_{\alpha\beta}] . \tag{3.59}$$

3.4 Strain Field

The Green-Lagrange strains and the Almansi strains are frequently used in numerical simulations, which are associated respectively with the second Piola-Kirchhoff stresses and the Cauchy stresses. The Green-Lagrange strains are referred to the undeformed configuration, while the Almansi strains are measured in the deformed configuration.

In problems of geometrically nonlinear analysis, the internal virtual work is defined as (see e.g. [9, 10])

$$\delta W_{\mathrm{int}} = \int_{\mathrm{V}} \sigma^{ij} \delta\varepsilon_{ij}\, \mathrm{dV} \tag{3.60}$$

where ε_{ij} and σ^{ij} denote the components of the Green-Lagrange strain tensor and the second Piola-Kirchhoff stress tensor, respectively. In such a way, the volume integral is referred to the undeformed configuration, which can be easily formulated. Due to this reason, the Green-Lagrange strains are mostly employed in large rotation theories.

The deformation gradient tensor $\boldsymbol{F}$, which maps the undeformed basis $\boldsymbol{g}^i$ into the deformed one $\bar{\boldsymbol{g}}_i$, is defined as

$$\boldsymbol{F} = \bar{\boldsymbol{g}}_i \otimes \boldsymbol{g}^i \,, \quad \boldsymbol{F}^{\mathrm{T}} = \boldsymbol{g}^i \otimes \bar{\boldsymbol{g}}_i \,. \tag{3.61}$$

With the help of the right Cauchy-Green tensor

$$\boldsymbol{C} = \boldsymbol{F}^{\mathrm{T}}\boldsymbol{F} = (\boldsymbol{g}^i \otimes \bar{\boldsymbol{g}}_i)(\bar{\boldsymbol{g}}_j \otimes \boldsymbol{g}^j) = \bar{g}_{ij}\, \boldsymbol{g}^i \otimes \boldsymbol{g}^j \,, \tag{3.62}$$

and the Riemannian metric tensor

$$\boldsymbol{G} = \boldsymbol{g}^i \otimes \boldsymbol{g}_i = \boldsymbol{g}_i \otimes \boldsymbol{g}^i = g_{ij}\boldsymbol{g}^i \otimes \boldsymbol{g}^j = g^{ij}\boldsymbol{g}_i \otimes \boldsymbol{g}_j \,, \tag{3.63}$$

the Green-Lagrange strain tensor is introduced and defined as (see books e.g. [11])

$$\boldsymbol{E} = \frac{1}{2}(\boldsymbol{C} - \boldsymbol{G}) \,, \tag{3.64}$$

Substituting Eqs. (3.62) and (3.63) into (3.64), one obtains the Green-Lagrange strain tensor

$$\boldsymbol{\varepsilon} = \frac{1}{2}(\bar{g}_{ij} - g_{ij})\, \boldsymbol{g}^i \otimes \boldsymbol{g}^j = \varepsilon_{ij}\, \boldsymbol{g}^i \otimes \boldsymbol{g}^j \,. \tag{3.65}$$

The components of the covariant metric tensor for an arbitrary point in the shell space associated with undeformed and deformed configurations can be constructed by base vectors at the mid-surface and their derivatives using Eq. (3.29). The components of the covariant metric tensor in the undeformed configuration can be obtained as

$$\begin{aligned}
g_{\alpha\beta} &= \boldsymbol{g}_\alpha \cdot \boldsymbol{g}_\beta = \boldsymbol{a}_\alpha \cdot \boldsymbol{a}_\beta + \Theta^3(\boldsymbol{n}_{,\alpha} \cdot \boldsymbol{a}_\beta + \boldsymbol{a}_\alpha \cdot \boldsymbol{n}_{,\beta}) + (\Theta^3)^2\, \boldsymbol{n}_{,\alpha} \cdot \boldsymbol{n}_{,\beta} \,, \\
g_{\alpha 3} &= \boldsymbol{g}_\alpha \cdot \boldsymbol{g}_3 = \boldsymbol{a}_\alpha \cdot \boldsymbol{n} + \Theta^3\, \boldsymbol{n}_{,\alpha} \cdot \boldsymbol{n} = 0 \,, \\
g_{33} &= \boldsymbol{g}_3 \cdot \boldsymbol{g}_3 = \boldsymbol{n} \cdot \boldsymbol{n} = 1 \,.
\end{aligned} \tag{3.66}$$

The components of the covariant metric tensor in the deformed configuration can be obtained in a similar way

$$\begin{aligned}
\bar{g}_{\alpha\beta} &= \bar{\boldsymbol{g}}_\alpha \cdot \bar{\boldsymbol{g}}_\beta = \bar{\boldsymbol{a}}_\alpha \cdot \bar{\boldsymbol{a}}_\beta + \Theta^3(\bar{\boldsymbol{a}}_{3,\alpha} \cdot \bar{\boldsymbol{a}}_\beta + \bar{\boldsymbol{a}}_\alpha \cdot \bar{\boldsymbol{a}}_{3,\beta}) + (\Theta^3)^2\, \bar{\boldsymbol{a}}_{3,\alpha} \cdot \bar{\boldsymbol{a}}_{3,\beta} \,, \\
\bar{g}_{\alpha 3} &= \bar{\boldsymbol{g}}_\alpha \cdot \bar{\boldsymbol{g}}_3 = \bar{\boldsymbol{a}}_\alpha \cdot \bar{\boldsymbol{a}}_3 + \Theta^3\, \bar{\boldsymbol{a}}_{3,\alpha} \cdot \bar{\boldsymbol{a}}_3 \,, \\
\bar{g}_{33} &= \bar{\boldsymbol{g}}_3 \cdot \bar{\boldsymbol{g}}_3 = \bar{\boldsymbol{a}}_3 \cdot \bar{\boldsymbol{a}}_3 \,.
\end{aligned} \tag{3.67}$$

Substituting the components of the covariant metric tensor in the shell space, given in (3.66) and (3.67), into the Green-Lagrange strain tensor, shown in (3.65), one obtains the in-plane, the transverse shear and the transverse normal components of the Green-Lagrange strain tensor in terms of the covariant base vectors at the mid-surface as (see Habip [12], who first developed the fully geometrically nonlinear strain-displacement relations based on FOSD hypothesis)

$$\varepsilon_{\alpha\beta} = \overset{0}{\varepsilon}_{\alpha\beta} + \Theta^3 \overset{1}{\varepsilon}_{\alpha\beta} + (\Theta^3)^2 \overset{2}{\varepsilon}_{\alpha\beta} , \tag{3.68}$$

$$\varepsilon_{\alpha 3} = \overset{0}{\varepsilon}_{\alpha 3} + \Theta^3 \overset{1}{\varepsilon}_{\alpha 3} , \tag{3.69}$$

$$\varepsilon_{33} = \overset{0}{\varepsilon}_{33} , \tag{3.70}$$

where the strain terms in the above equations are

$$2\overset{0}{\varepsilon}_{\alpha\beta} = \bar{\boldsymbol{a}}_\alpha \cdot \bar{\boldsymbol{a}}_\beta - \boldsymbol{a}_\alpha \cdot \boldsymbol{a}_\beta , \tag{3.71}$$

$$2\overset{1}{\varepsilon}_{\alpha\beta} = \bar{\boldsymbol{a}}_\alpha \cdot \bar{\boldsymbol{a}}_{3,\beta} + \bar{\boldsymbol{a}}_{3,\alpha} \cdot \bar{\boldsymbol{a}}_\beta - \boldsymbol{a}_\alpha \cdot \boldsymbol{a}_{3,\beta} - \boldsymbol{a}_{3,\alpha} \cdot \boldsymbol{a}_\beta , \tag{3.72}$$

$$2\overset{2}{\varepsilon}_{\alpha\beta} = \bar{\boldsymbol{a}}_{3,\alpha} \cdot \bar{\boldsymbol{a}}_{3,\beta} - \boldsymbol{a}_{3,\alpha} \cdot \boldsymbol{a}_{3,\beta} , \tag{3.73}$$

$$2\overset{0}{\varepsilon}_{\alpha 3} = \bar{\boldsymbol{a}}_\alpha \cdot \bar{\boldsymbol{a}}_3 , \tag{3.74}$$

$$2\overset{1}{\varepsilon}_{\alpha 3} = \bar{\boldsymbol{a}}_{3,\alpha} \cdot \bar{\boldsymbol{a}}_3 , \tag{3.75}$$

$$2\overset{0}{\varepsilon}_{33} = \bar{\boldsymbol{a}}_3 \cdot \bar{\boldsymbol{a}}_3 - 1 . \tag{3.76}$$

Here, $(\overset{0}{\varepsilon}_{11}, \overset{0}{\varepsilon}_{22})$ represent the in-plane longitudinal strains, $(\overset{0}{\varepsilon}_{12}, \overset{0}{\varepsilon}_{21})$ are the in-plane shear strains, $(\overset{1}{\varepsilon}_{11}, \overset{1}{\varepsilon}_{22})$ denote the bending strains, $(\overset{1}{\varepsilon}_{12}, \overset{1}{\varepsilon}_{21})$ are the torsional strains, $(\overset{0}{\varepsilon}_{13}, \overset{0}{\varepsilon}_{23})$ are the transverse shear strains, $\overset{0}{\varepsilon}_{33}$ denotes the transverse normal strain. Additionally, $\overset{2}{\varepsilon}_{\alpha\beta}$, $\overset{1}{\varepsilon}_{\alpha 3}$ are corrections respectively for the in-plane and shear strains.

Considering the relations given in (3.34) and (3.35), the Green-Lagrange strain components can be obtained in terms of the base vectors and displacement vectors in the undeformed configuration as

$$2\overset{0}{\varepsilon}_{\alpha\beta} = \boldsymbol{a}_\alpha \cdot \overset{0}{\boldsymbol{u}}_{,\beta} + \overset{0}{\boldsymbol{u}}_{,\alpha} \cdot \boldsymbol{a}_\beta + \overset{0}{\boldsymbol{u}}_{,\alpha} \cdot \overset{0}{\boldsymbol{u}}_{,\beta} , \tag{3.77}$$

$$\begin{aligned} 2\overset{1}{\varepsilon}_{\alpha\beta} = {} & \boldsymbol{a}_\alpha \cdot \overset{1}{\boldsymbol{u}}_{,\beta} + \overset{0}{\boldsymbol{u}}_{,\alpha} \cdot \overset{1}{\boldsymbol{u}}_{,\beta} + \overset{0}{\boldsymbol{u}}_{,\alpha} \cdot \boldsymbol{n}_{,\beta} \\ & + \overset{1}{\boldsymbol{u}}_{,\alpha} \cdot \boldsymbol{a}_\beta + \overset{1}{\boldsymbol{u}}_{,\alpha} \cdot \overset{0}{\boldsymbol{u}}_{,\beta} + \boldsymbol{n}_{,\alpha} \cdot \overset{0}{\boldsymbol{u}}_{,\beta} , \end{aligned} \tag{3.78}$$

$$2\overset{2}{\varepsilon}_{\alpha\beta} = \overset{1}{\boldsymbol{u}}_{,\alpha} \cdot \overset{1}{\boldsymbol{u}}_{,\beta} + \overset{1}{\boldsymbol{u}}_{,\alpha} \cdot \boldsymbol{n}_{,\beta} + \boldsymbol{n}_{,\alpha} \cdot \overset{1}{\boldsymbol{u}}_{,\beta} , \tag{3.79}$$

$$2\overset{0}{\varepsilon}_{\alpha 3} = \boldsymbol{a}_\alpha \cdot \overset{1}{\boldsymbol{u}} + \boldsymbol{a}_\alpha \cdot \boldsymbol{n} + \overset{0}{\boldsymbol{u}}_{,\alpha} \cdot \overset{1}{\boldsymbol{u}} + \overset{0}{\boldsymbol{u}}_{,\alpha} \cdot \boldsymbol{n} , \tag{3.80}$$

$$2\overset{1}{\varepsilon}_{\alpha 3} = \overset{1}{\boldsymbol{u}}_{,\alpha} \cdot \overset{1}{\boldsymbol{u}} + \overset{1}{\boldsymbol{u}}_{,\alpha} \cdot \boldsymbol{n} + \boldsymbol{n}_{,\alpha} \cdot \overset{1}{\boldsymbol{u}} + \boldsymbol{n}_{,\alpha} \cdot \boldsymbol{n} , \tag{3.81}$$

$$2\overset{0}{\varepsilon}_{33} = \overset{1}{\boldsymbol{u}} \cdot \overset{1}{\boldsymbol{u}} + \overset{1}{\boldsymbol{u}} \cdot \boldsymbol{n} + \boldsymbol{n} \cdot \overset{1}{\boldsymbol{u}} + \boldsymbol{n} \cdot \boldsymbol{n} - 1 . \tag{3.82}$$

Substituting Eqs. (3.36), (3.37) and (3.48) into (3.77)–(3.82) yields the strain-displacement relations in terms of six parameters as

$$2\overset{0}{\varepsilon}_{\alpha\beta} = \overset{0}{\varphi}_{\alpha\beta} + \overset{0}{\varphi}_{\beta\alpha} + \underline{\overset{0}{\varphi}_{3\alpha}\overset{0}{\varphi}_{3\beta}} + \underline{\underline{\overset{0}{\varphi}{}^{\delta}_{\alpha}\overset{0}{\varphi}_{\delta\beta}}}\,, \tag{3.83}$$

$$\begin{aligned}2\overset{1}{\varepsilon}_{\alpha\beta} = {} & \overset{1}{\varphi}_{\alpha\beta} - b^{\lambda}_{\beta}\overset{0}{\varphi}_{\lambda\alpha} + \overset{1}{\varphi}_{\beta\alpha} - b^{\delta}_{\alpha}\overset{0}{\varphi}_{\delta\beta} \\ & + \underline{\overset{0}{\varphi}_{3\alpha}\overset{1}{\varphi}_{3\beta} + \overset{1}{\varphi}_{3\alpha}\overset{0}{\varphi}_{3\beta}} + \underline{\underline{\overset{0}{\varphi}{}^{\delta}_{\alpha}\overset{1}{\varphi}_{\delta\beta} + \overset{1}{\varphi}{}^{\delta}_{\alpha}\overset{0}{\varphi}_{\delta\beta}}}\,,\end{aligned} \tag{3.84}$$

$$2\overset{2}{\varepsilon}_{\alpha\beta} = -b^{\lambda}_{\beta}\overset{1}{\varphi}_{\lambda\alpha} - b^{\delta}_{\alpha}\overset{1}{\varphi}_{\delta\beta} + \underline{\overset{1}{\varphi}_{3\alpha}\overset{1}{\varphi}_{3\beta}} + \underline{\underline{\overset{1}{\varphi}{}^{\delta}_{\alpha}\overset{1}{\varphi}_{\delta\beta}}}\,, \tag{3.85}$$

$$2\overset{0}{\varepsilon}_{\alpha 3} = \overset{1}{v}_{\alpha} + \overset{0}{\varphi}_{3\alpha} + \underline{\overset{0}{\varphi}{}^{\delta}_{\alpha}\overset{1}{v}_{\delta}} + \underline{\underline{\overset{0}{\varphi}_{3\alpha}\overset{1}{v}_{3}}}\,, \tag{3.86}$$

$$2\overset{1}{\varepsilon}_{\alpha 3} = \overset{1}{\varphi}_{3\alpha} - b^{\delta}_{\alpha}\overset{1}{v}_{\delta} + \underline{\overset{1}{\varphi}{}^{\delta}_{\alpha}\overset{1}{v}_{\delta}} + \underline{\underline{\overset{1}{\varphi}_{3\alpha}\overset{1}{v}_{3}}}\,, \tag{3.87}$$

$$2\overset{0}{\varepsilon}_{33} = 2\,\overset{1}{v}_{3} + a^{\lambda\delta}\overset{1}{v}_{\lambda}\overset{1}{v}_{\delta} + (\overset{1}{v}_{3})^{2}\,. \tag{3.88}$$

3.5 Shell Theories

In Sect. 3.4, the fully geometrically nonlinear strain-displacement relations are discussed. In the framework of FOSD hypothesis, six parameters are introduced. For the simplified nonlinear or linear strain-displacement relations, the six parameters are reduced to five parameters. The definitions of linear and nonlinear shell theories associated with the number of parameters are listed in Table 3.2. Furthermore, it is assumed that the shell director in thin-walled structures is inextensible, which leads to $\overset{0}{\varepsilon}_{33} = 0$.

The physical meanings of the six independent kinematic parameters ($\overset{0}{v}_{\alpha}$, $\overset{0}{v}_{3}$, $\overset{1}{v}_{\alpha}$, $\overset{1}{v}_{3}$) in the fully geometrically nonlinear relations are not clear. These six parameters are usually expressed by nodal DOFs which have specific physical meanings. It is

Table 3.2 List of nonlinear shell theories based on FOSD hypothesis

Theory	Specification	Parameters
LRT56	Large rotation shell theory with six parameters expressed by five nodal DOFs	$\overset{0}{v}_{\alpha}$, $\overset{0}{v}_{3}$, $\overset{1}{v}_{\alpha}$, $\overset{1}{v}_{3}$
LRT5	Fully geometrically nonlinear shell theory with five parameters	$\overset{0}{v}_{\alpha}$, $\overset{0}{v}_{3}$, $\overset{1}{v}_{\alpha}$
MRT5	Moderate rotation shell theory with five parameters	
RVK5	Refined von Kármán type nonlinear shell theory with five parameters	
LIN5	Geometrically linear shell theory with five parameters	$\overset{0}{v}_{\alpha}$, $\overset{0}{v}_{3}$, $\overset{1}{v}_{\alpha}$

Table 3.3 Strain-displacement relations for various shell theories

Strain	Strain-displacement relation	Theory
$2\overset{0}{\varepsilon}_{\alpha\beta} =$	$\overset{0}{\varphi}_{\alpha\beta} + \overset{0}{\varphi}_{\beta\alpha} + \overset{0}{\varphi}_{3\alpha}\overset{0}{\varphi}_{3\beta} + \overset{0}{\varphi}{}^{\delta}_{\alpha}\overset{0}{\varphi}_{\delta\beta}$	LRT56
		LRT5
	$\overset{0}{\varphi}_{\alpha\beta} + \overset{0}{\varphi}_{\beta\alpha} + \overset{0}{\varphi}_{3\alpha}\overset{0}{\varphi}_{3\beta}$	MRT5
	$\overset{0}{\varphi}_{\alpha\beta} + \overset{0}{\varphi}_{\beta\alpha} + \overset{0}{v}_{3,\alpha}\overset{0}{v}_{3,\beta}$	RVK5
	$\overset{0}{\varphi}_{\alpha\beta} + \overset{0}{\varphi}_{\beta\alpha}$	LIN5
$2\overset{1}{\varepsilon}_{\alpha\beta} =$	$\overset{1}{\varphi}_{\alpha\beta} - b^{\lambda}_{\beta}\overset{0}{\varphi}_{\lambda\alpha} + \overset{1}{\varphi}_{\beta\alpha} - b^{\delta}_{\alpha}\overset{0}{\varphi}_{\delta\beta} +$ $\overset{0}{\varphi}_{3\alpha}\overset{1}{\varphi}_{3\beta} + \overset{1}{\varphi}_{3\alpha}\overset{0}{\varphi}_{3\beta} + \overset{0}{\varphi}{}^{\delta}_{\alpha}\overset{1}{\varphi}_{\delta\beta} + \overset{1}{\varphi}{}^{\delta}_{\alpha}\overset{0}{\varphi}_{\delta\beta}$	LRT56
		LRT5
	$\overset{1}{\varphi}_{\alpha\beta} - b^{\lambda}_{\beta}\overset{0}{\varphi}_{\lambda\alpha} + \overset{1}{\varphi}_{\beta\alpha} - b^{\delta}_{\alpha}\overset{0}{\varphi}_{\delta\beta} +$ $\overset{0}{\varphi}_{3\alpha}\overset{1}{\varphi}_{3\beta} + \overset{1}{\varphi}_{3\alpha}\overset{0}{\varphi}_{3\beta}$	MRT5
	$\overset{1}{\varphi}_{\alpha\beta} - b^{\lambda}_{\beta}\overset{0}{\varphi}_{\lambda\alpha} + \overset{1}{\varphi}_{\beta\alpha} - b^{\delta}_{\alpha}\overset{0}{\varphi}_{\delta\beta}$	RVK5
		LIN5
$2\overset{2}{\varepsilon}_{\alpha\beta} =$	$-b^{\lambda}_{\beta}\overset{1}{\varphi}_{\lambda\alpha} - b^{\delta}_{\alpha}\overset{1}{\varphi}_{\delta\beta} + \overset{1}{\varphi}_{3\alpha}\overset{1}{\varphi}_{3\beta} + \overset{1}{\varphi}{}^{\delta}_{\alpha}\overset{1}{\varphi}_{\delta\beta}$	LRT56
		LRT5
	$-b^{\lambda}_{\beta}\overset{1}{\varphi}_{\lambda\alpha} - b^{\delta}_{\alpha}\overset{1}{\varphi}_{\delta\beta} + \overset{1}{\varphi}_{3\alpha}\overset{1}{\varphi}_{3\beta}$	MRT5
	$-b^{\lambda}_{\beta}\overset{1}{\varphi}_{\lambda\alpha} - b^{\delta}_{\alpha}\overset{1}{\varphi}_{\delta\beta}$	RVK5
		LIN5
$2\overset{0}{\varepsilon}_{\alpha 3} =$	$\overset{1}{v}_{\alpha} + \overset{0}{\varphi}_{3\alpha} + \overset{0}{\varphi}{}^{\delta}_{\alpha}\overset{1}{v}_{\delta} + \overset{0}{\varphi}_{3\alpha}\overset{1}{v}_{3}$	LRT56
	$\overset{1}{v}_{\alpha} + \overset{0}{\varphi}_{3\alpha} + \overset{0}{\varphi}{}^{\delta}_{\alpha}\overset{1}{v}_{\delta}$	LRT5
		MRT5
	$\overset{1}{v}_{\alpha} + \overset{0}{\varphi}_{3\alpha}$	RVK5
		LIN5
$2\overset{1}{\varepsilon}_{\alpha 3} =$	$\overset{1}{\varphi}_{3\alpha} - b^{\delta}_{\alpha}\overset{1}{v}_{\delta} + \overset{1}{\varphi}{}^{\delta}_{\alpha}\overset{1}{v}_{\delta} + \overset{1}{\varphi}_{3\alpha}\overset{1}{v}_{3}$	LRT56
	$\overset{1}{\varphi}_{3\alpha} - b^{\delta}_{\alpha}\overset{1}{v}_{\delta} + \overset{1}{\varphi}{}^{\delta}_{\alpha}\overset{1}{v}_{\delta}$	LRT5
		MRT5
	$\overset{1}{\varphi}_{3\alpha} - b^{\delta}_{\alpha}\overset{1}{v}_{\delta}$	RVK5
		LIN5
$2\overset{0}{\varepsilon}_{33} =$	0	LRT56
		LRT5
		MRT5
		RVK5
		LIN5

Table 3.4 The expressions of the abbreviations for various shell theories

Theory	$\overset{n}{\varphi}_{\lambda\alpha} =$	$\overset{n}{\varphi}_{3\alpha} =$
LRT56	$\overset{0}{v}_{\lambda,\alpha} - \Gamma^{\delta}_{\lambda\alpha}\overset{0}{v}_{\delta} - b_{\lambda\alpha}\overset{0}{v}_3$	$\overset{0}{v}_{3,\alpha} + b^{\delta}_{\alpha}\overset{0}{v}_{\delta}$
	$\overset{1}{v}_{\lambda,\alpha} - \Gamma^{\delta}_{\lambda\alpha}\overset{1}{v}_{\delta} - b_{\lambda\alpha}\overset{1}{v}_3$	$\overset{1}{v}_{3,\alpha} + b^{\delta}_{\alpha}\overset{1}{v}_{\delta}$
LRT5, MRT5, RVK5, LIN5	$\overset{0}{v}_{\lambda,\alpha} - \Gamma^{\delta}_{\lambda\alpha}\overset{0}{v}_{\delta} - b_{\lambda\alpha}\overset{0}{v}_3$	$\overset{0}{v}_{3,\alpha} + b^{\delta}_{\alpha}\overset{0}{v}_{\delta}$
	$\overset{1}{v}_{\lambda,\alpha} - \Gamma^{\delta}_{\lambda\alpha}\overset{1}{v}_{\delta}$	$b^{\delta}_{\alpha}\overset{1}{v}_{\delta}$

assumed that the rotation about the thickness axis is not applicable in thin-walled laminated smart structures, resulting in only five nodal DOFs. Considering fully geometrically nonlinear strain-displacement relations, in which the six parameters are expressed by five nodal DOFs (see Chap. 5), the resulting theory is abbreviated as LRT56 theory (see [13–16]). For simplified nonlinear theories, the sixth parameter $\overset{1}{v}_3$ is usually Neglected, due to the assumption of small or moderate rotations. The five parameters for the simplified nonlinear theories are respectively equal to five DOFs (more details refers to Chap. 5). Using five parameters with consideration of full geometric nonlinearities one obtains a theory abbreviated as LRT5 [13–16]. Further removing the nonlinear strain-displacement terms marked by double lines in (3.83)–(3.88) yields the moderate rotation theory (MRT5), which was earlier developed by Schmidt and Reddy [17], (see also [13, 14, 16, 18–22]). Again dropping more nonlinear terms one obtains the simplest nonlinear theory, refined von Kármán type nonlinear theory. The refined von Kármán type nonlinear theory retains only the nonlinear terms containing the squares and products of derivatives of the transverse deflection in the in-plane longitudinal and shear strain components, abbreviated as RVK5 [15, 16]. Dropping all the nonlinear terms marked by both single and double lines results in linear theory with five parameters, which is shorted as LIN5.

The strain-displacement relations for various shell theories mentioned above can be obtained as shown in Table 3.3, by using the abbreviations listed in Table 3.4.

3.6 Normalization

From Eqs. (3.36), (3.37) and (3.65), it can be seen that the components of the displacement and strain tensors are associated with the base vectors which are not necessarily unit vectors. Therefore, normalized components of the displacement and strain vectors with physical meanings should be introduced, which are obtained by normalization. The displacement vector is defined with respect to the mid-surface contravariant basis as

$$\overset{n}{\boldsymbol{u}} = \overset{n}{v}_i\, \boldsymbol{a}^i = \overset{n}{v}_1\, \boldsymbol{a}^1 + \overset{n}{v}_2\, \boldsymbol{a}^2 + \overset{n}{v}_3\, \boldsymbol{a}^3 \,. \tag{3.89}$$

It can also be expressed in the corresponding contravariant basis, but with unit Euclidean length, as

$$\overset{n}{\boldsymbol{u}} = \overset{\hat{n}}{v}_1 \hat{\boldsymbol{a}}^1 + \overset{\hat{n}}{v}_2 \hat{\boldsymbol{a}}^2 + \overset{\hat{n}}{v}_3 \hat{\boldsymbol{a}}^3 , \tag{3.90}$$

where, $\overset{\hat{n}}{v}_i$ denote the physical quantity of $\overset{n}{v}_i$, and $\hat{\boldsymbol{a}}^i = \frac{\boldsymbol{a}^i}{\|\boldsymbol{a}^i\|}$ represents the normalized vectors of $\boldsymbol{a}^i$. From Eqs. (3.89) and (3.90), one can easily obtain

$$\overset{n}{v}_i = \frac{\overset{\hat{n}}{v}_i}{\|\boldsymbol{a}^i\|} . \tag{3.91}$$

Analogously, the physical components of the Green-Lagrange strain tensor, which is a second-order tensor expressed by the contravariant basis $\boldsymbol{g}^i \otimes \boldsymbol{g}^j$ in the shell space, can be calculated by the same procedure as

$$\boldsymbol{\varepsilon} = \varepsilon_{ij}\, \boldsymbol{g}^i \otimes \boldsymbol{g}^j = \hat{\varepsilon}_{ij}\, \hat{\boldsymbol{g}}^i \otimes \hat{\boldsymbol{g}}^j . \tag{3.92}$$

Here again $\hat{\boldsymbol{g}}^i = \frac{\boldsymbol{g}^i}{\|\boldsymbol{g}^i\|}$ represents the normalized vector of $\boldsymbol{g}^i$, such that the physical components of the strain tensor are

$$\hat{\varepsilon}_{ij} = \|\boldsymbol{g}^i\|\|\boldsymbol{g}^j\|\varepsilon_{ij} . \tag{3.93}$$

3.7 Summary

This chapter deduced fully and simplified geometrically nonlinear strain-displacement relations based on FOSD hypothesis for various nonlinear shell theories. The differences between each nonlinear shell theory were analyzed and strengthened.

References

1. H.S. Tzou, M. Gadre, Theoretical analysis of a multi-layered thin shell coupled with piezoelectric shell actuators for distributed vibration controls. J. Sound Vib. **132**, 433–450 (1989)
2. H. Kioua, S. Mirza, Piezoelectric induced bending and twisting of laminated composite shallow shells. Smart Mater. Struct. **9**, 476–484 (2000)
3. K.Y. Lam, X.Q. Peng, G.R. Liu, J.N. Reddy, A finite-element model for piezoelectric composite laminates. Smart Mater. Struct. **6**, 583–591 (1997)
4. G.R. Liu, X.Q. Peng, K.Y. Lam, J. Tani, Vibration control simulation of laminated composite plates with integrated piezoelectrics. J. Sound Vib. **220**, 827–846 (1999)
5. J.M.S. Moita, I.F.P. Correia, C.M. Soares, C.A.M. Soares, Active control of adaptive laminated structures with bonded piezoelectric sensors and actuators. Comput. Struct. **82**, 1349–1358 (2004)

6. J.M.S. Moita, V.M.F. Correia, P.G. Martins, C.M.M. Soares, C.A.M. Soares, Optimal design in vibration control of adaptive structures using a simulated annealing algorithm. Compos. Struct. **75**, 79–87 (2006)
7. H.Y. Zhang, Y.P. Shen, Vibration suppression of laminated plates with 1–3 piezoelectric fiber-reinforced composite layers equipped with integrated electrodes. Compos. Struct. **79**, 220–228 (2007)
8. R.D. Mindlin, Forced thickness-shear and flexural vibrations of piezoelectric crystal plates. J. Appl. Phys. **23**, 83–88 (1952)
9. A.E. Green, W. Zerna, *Theoretical Elasticity*, 2nd edn. (Clarendon Press, Oxford, 1968)
10. A.E. Green, J.E. Adkins, *Large Elastic Deformations*, 2nd edn. (Clarendon Press, Oxford, 1970)
11. Y. Basar, D. Weichert, *Nonlinear Continuum Mechanics of Solids: Fundamental mathematical and physical concepts* (Springer, Berlin Germany, 1999)
12. L.M. Habip, Theory of elastic shells in the reference state. Ingenieur-Archiv **34**, 228–237 (1965)
13. S.Q. Zhang, R. Schmidt, Large rotation theory for static analysis of composite and piezoelectric laminated thin-walled structures. Thin-Walled Struct. **78**, 16–25 (2014)
14. S.Q. Zhang, R. Schmidt, Large rotation FE transient analysis of piezolaminated thin-walled smart structures. Smart Mater. Struct. **22**, 105025 (2013)
15. I. Kreja, R. Schmidt, Large rotations in first-order shear deformation FE analysis of laminated shells. Int. J. Non-Linear Mech. **41**, 101–123 (2006)
16. I. Kreja, *Geometrically non-linear analysis of layered composite plates and shells*. Habilitation Thesis, Published as Monografie 83, Politechnika Gdańska (2007)
17. R. Schmidt, J.N. Reddy, A refined small strain and moderate rotation theory of elastic anisotropic shells. J. Appl. Mech. **55**, 611–617 (1988)
18. L. Librescu, R. Schmidt, Refined theories of elastic anisotropic shells accounting for small strains and moderate rotations. Int. J. Non-Linear Mech. **23**, 217–229 (1988)
19. R. Schmidt, D. Weichert, A refined theory of elastic-plastic shells at moderate rotations. ZAMM · Z. Angew. Math. Mech. **69**, 11–21 (1989)
20. A.F. Palmerio, J.N. Reddy, R. Schmidt, On a moderate rotation theory of laminated anisotropic shells - part 1: theory. Int. J. Non-Linear Mech. **25**, 687–700 (1990)
21. A.F. Palmerio, J.N. Reddy, R. Schmidt, On a moderate rotation theory of laminated anisotropic shells - part 2: finite element analysis. Int. J. Non-Linear Mech. **25**, 701–714 (1990)
22. I. Kreja, R. Schmidt, J.N. Reddy, Finite elements based on a first-order shear deformation moderate rotation shell theory with applications to the analysis of composite structures. Int. J. Non-Linear Mech. **32**, 1123–1142 (1996)

Chapter 4
Nonlinear Constitutive Relations

Abstract This chapter deals with constitutive relations for piezoelectric materials with isotropy or orthotropy. Firstly, piezoelectricity is introduced for brief understanding of piezoelectric materials. To deep understand the basic principles of piezo materials, the fundamental theory of piezoelectricity is discussed in the three-dimensional case. In order to deal with fiber based piezo materials or fiber reinforced composites, coordinate transformation law between fiber coordinates (material coordinates) and convective coordinates (structural coordinates) is introduced. Afterwards, the constitutive relations for two typical configurations of macro-fiber composite piezo materials are developed, where multi-layered structures are considered. Finally, an electro-mechanically coupled nonlinear constitutive relations for piezoelectric with either isotropy or orthotropy are constructed.

4.1 Piezoelectricity

4.1.1 History of Piezoelectricity

The phenomenon piezoelectricity was first discovered by the brothers Pierre Curie and Jacques Curie in 1880. They found that some crystals, e.g. quartz, Rochelle salt, cane sugar, etc., will produce positive or negative electric charges under a compressive load. The amount of electric charges were found to be proportional to the applied compressive load. This effect of generation of electric charges because of compressive load is referred to as "direct effect". In contrast to "direct effect", piezoelectric material has "converse effect". The converse effect sometimes is referred to as reciprocal or inverse effect, which describes that an additional strain or deformation will be caused by an electric field. Again, the induced strain is proportional to the applied electric charges. The converse effect of piezoelectric material was first mathematically proved through fundamental thermo-dynamic principles by Gabriel Lippmann in 1881. In the same year, the complete reversibility of electroelastic deformations in piezoelectric crystals was experimentally demonstrated by the Curie brothers.

S.-Q. Zhang, *Nonlinear Analysis of Thin-Walled Smart Structures*, Springer Tracts in Mechanical Engineering, https://doi.org/10.1007/978-981-15-9857-9_4

After the discovery of piezoelectricity, due to the high operational frequency (in the range of one megaHertz) of quartz and Rochelle-salt plates, many applications were implemented ranging from radio transmitter, underwater detection, pressure measurement to many kinds of electrical measurements, microphones, and accelerometers. Around the Second World War, polycrystalline piezoceramic materials was discovered, which has high dielectric constants and could be manufactured in high volumes. The low piezoelectric effect of nature materials limits the applications until ferroelectric material was found during the Second World War. The man-made ferroelectrics exhibited piezoelectric effects many times higher than those found in natural materials. In 1940s, Arthur von Hippel and coworkers at MIT discovered barium titanate ($BaTiO_3$) that has the capability of repolarization under a high electric field. However, the Curie Temperature takes only 120°, which means that the piezoelectric effect disappears when the temperature is above 120°. In 1950s, with the discovery of piezoelectric effects in lead metaniobate ($PbNb_2O_6$) and lead zirconate titanate [$Pb(Ti,Zr)O_6$], the Curie Temperature increases to 250°. Different from ceramic, a soft piezoelectric material, polymer polyvinylidenefluoride (PVDF), was discovered by Kawai [1]. Due to the flexibility, PVDF is frequently manufactured in thin films, which is easy to fit curved geometries. However, the low stiffness makes the material usually used as sensors.

There are several practical limitations in the applications of piezoceramic materials, for example the brittle nature of ceramics which makes them susceptible to fracture during handling and bonding procedures, and their extremely limited ability to fit with curved surfaces [2]. Even though the PVDF material is soft and flexible, but with low stiffness, which are only used for sensors. To overcome the limitations existing in conventional piezoelectric materials, piezo composite materials were proposed and developed by some researchers in the 1990s. The first type of piezo composite is referred to as 1-3 composite invented at the Fraunhofer Research Facility in Germany [2]. The second one is an active fiber composite (AFC) initially developed by MIT, which were the first composite actuators primarily used on structural actuation [2]. The third one is a macro-fiber composite (MFC) proposed by NASA Langley Research Center [3] in 1999. The flexible nature of MFC allows the material conforming to a curved surface easily. Additionally, an MFC patch even has larger actuation forces than a PZT patch, since the d_{33} effect dominates the actuation mode in MFCs. For more detailed information of active piezoelectric fiber composites, we refer to [4–6].

4.1.2 Piezoelectric Effects

The raw piezoceramics illustrate electrically neutral, without piezoelectric effect. They need to be polarized by applying strong electric field. In most cases the piezoelectric materials are also ferroelectric, the piezoelectric phase can be transformed to a symmetric non-piezoelectric state at a certain high temperature, which here refers to the Curie temperature, as shown in Fig. 4.1. The ion Ti^{4+} in the center will be

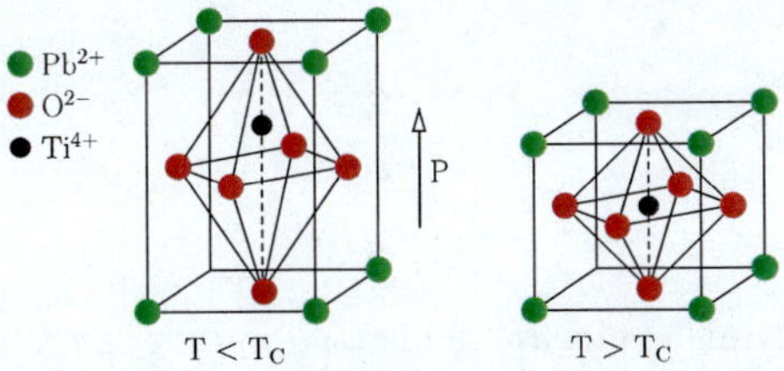

Fig. 4.1 The configurations of $PbTiO_3$ crystalline structure

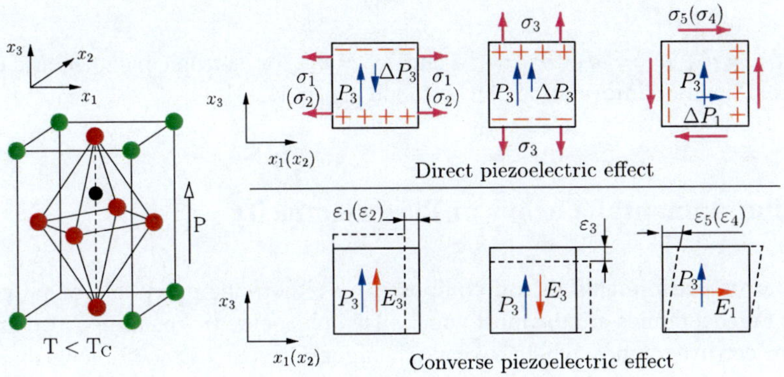

Fig. 4.2 The direct and converse effects of piezoelectric material

shifted to one side of the crystalline structure when the temperature is below the Curie point. As a consequence, the center of the positive electric charges of the unit cell is different from that of the negative ones. The crystal is then called polarized.

The piezoelectric material has two effects, namely the direct and converse effects, which are shown in Fig. 4.2. Applying a stress in direction x_1, it will decrease the distance between the ion of titanium and the geometric center of the unit cell. This can be understood as an additionally generated *polarization*, which results in extra electric charges due to the stresses. Similarly, applying a normal stress σ_{33} or shear stress σ_{13}, one produces electric charges as well. Those phenomena are called *direct piezoelectric effect*, which can be expressed separately as

$$\begin{aligned} \Delta P_1 &= d_{15}\sigma_5 \,, \\ \Delta P_2 &= d_{24}\sigma_4 \,, \\ \Delta P_3 &= d_{31}\sigma_1 + d_{32}\sigma_2 + d_{33}\sigma_3 \,, \end{aligned} \tag{4.1}$$

where ΔP_i denotes the extra polarization in x_i direction.

In an analogous way, the physical meaning of the *converse* piezoelectric effect can be observed. Applying an electric field along the polarization direction will move the ion of titanium off the center in x_3 direction. This will result in stretching the cell along direction x_3 and squeezing along direction x_1 and x_2, which yields additional strains given as

$$\begin{aligned}\varepsilon_1 &= d_{13}E_3 ,\\ \varepsilon_2 &= d_{23}E_3 ,\\ \varepsilon_3 &= d_{33}E_3 .\end{aligned} \tag{4.2}$$

In the same way, applying an electric field along x_1 or x_2 direction yields additional shear strains as

$$\begin{aligned}\varepsilon_4 &= d_{42}E_2 ,\\ \varepsilon_5 &= d_{51}E_1 .\end{aligned} \tag{4.3}$$

Here, $d_{13} = d_{31}, d_{23} = d_{32}, d_{42} = d_{24}$ and $d_{51} = d_{15}$ for isotropic piezoelectric material. More detailed information can be found e.g. in [7, 8].

4.2 Fundamental Theory of Piezoelectricity

Most piezoelectric materials are composed by either single crystals or polycrystalline. Piezoceramics are the most widely used piezoelectric materials, also known as ferroelectric ceramics, which have much larger piezoelectric coefficients than natural crystals. In the original unprocessed form, these materials have no piezoelectric properties. However, the materials can be polarized by applying a strong electric field

Piezoceramics can be considered as isotropic material. Using the assumptions of small strains and weak electric field for piezoelectric patches or layers, the constitutive relations can be expressed as [9]

$$\varepsilon_{ij} = s_{ijkl}\sigma^{kl} + d_{ij\cdot}^{\ \ m}E_m , \tag{4.4}$$

$$D^m = d^m_{\cdot kl}\sigma^{kl} + \epsilon^{mn}E_n . \tag{4.5}$$

Here ε_{ij} is the strain tensor, σ^{kl} is the stress tensor, s_{ijkl} is the compliance tensor, $d^m_{\cdot kl}$ is the mixed piezoelectric constants tensor ($d_{ij\cdot}^{\ \ m}$ is the transposed tensor), ϵ^{mn} is the dielectric constant tensor, E_m is the electric field tensor, and D^m is the electric displacement tensor. Furthermore, the second-order strain and stress tensors are organized as

$$[\sigma^{ij}] = \begin{bmatrix}\sigma^{11} & \sigma^{12} & \sigma^{13}\\ \sigma^{21} & \sigma^{22} & \sigma^{23}\\ \sigma^{31} & \sigma^{32} & \sigma^{33}\end{bmatrix} , \quad [\varepsilon_{ij}] = \begin{bmatrix}\varepsilon_{11} & \varepsilon_{12} & \varepsilon_{13}\\ \varepsilon_{21} & \varepsilon_{22} & \varepsilon_{23}\\ \varepsilon_{31} & \varepsilon_{32} & \varepsilon_{33}\end{bmatrix} . \tag{4.6}$$

Due to the symmetry of the stress and strain tensors, $\sigma^{ij} = \sigma^{ji}$ and $\varepsilon_{ij} = \varepsilon_{ji}$, the Voigt notations are introduced to describe the second-order strain and stress tensors in vector form, which are defined as listed in Table 4.1. In such a way, the strains and stresses can be arranged in vector form as

Table 4.1 Voigt notation

ij or *kl*	*p* or *q*
11	1
22	2
33	3
23 or 32	4
13 or 31	5
12 or 21	6

$$\boldsymbol{\sigma} = \begin{Bmatrix} \sigma^{11} \\ \sigma^{22} \\ \sigma^{33} \\ \sigma^{23} \\ \sigma^{13} \\ \sigma^{12} \end{Bmatrix} = \begin{Bmatrix} \sigma_1 \\ \sigma_2 \\ \sigma_3 \\ \sigma_4 \\ \sigma_5 \\ \sigma_6 \end{Bmatrix}, \quad \boldsymbol{\varepsilon} = \begin{Bmatrix} \varepsilon_{11} \\ \varepsilon_{22} \\ \varepsilon_{33} \\ 2\varepsilon_{23} \\ 2\varepsilon_{13} \\ 2\varepsilon_{12} \end{Bmatrix} = \begin{Bmatrix} \varepsilon_1 \\ \varepsilon_2 \\ \varepsilon_3 \\ 2\varepsilon_4 \\ 2\varepsilon_5 \\ 2\varepsilon_6 \end{Bmatrix}. \tag{4.7}$$

In the piezoelectric integrated smart structures, two typical types of material are usually considered, namely pure metal material and fiber reinforced composite material. The former one can be described by isotropic material model, while the latter one can be represented by orthotropic material model. Additionally using the Voigt notations, the components of the fourth-order compliance constant tensor in (4.4) can be arranged in matrix form as

$$[s_{ijkl}] = \begin{bmatrix} s_{1111} & s_{1122} & s_{1123} & 0 & 0 & 0 \\ s_{1122} & s_{2222} & s_{2233} & 0 & 0 & 0 \\ s_{1133} & s_{2233} & s_{3333} & 0 & 0 & 0 \\ 0 & 0 & 0 & s_{2323} & 0 & 0 \\ 0 & 0 & 0 & 0 & s_{1313} & 0 \\ 0 & 0 & 0 & 0 & 0 & s_{1212} \end{bmatrix}. \tag{4.8}$$

In a more general case of orthotropic materials, the components in (4.8) are given by

$$\begin{aligned} s_{1111} &= \frac{1}{Y_1}, \quad & s_{2222} &= \frac{1}{Y_2}, \quad & s_{3333} &= \frac{1}{Y_3}, \\ s_{1122} &= -\frac{\nu_{12}}{Y_1}, \quad & s_{1133} &= -\frac{\nu_{13}}{Y_1}, \quad & s_{2233} &= -\frac{\nu_{23}}{Y_2}, \\ s_{2323} &= \frac{1}{G_{23}}, \quad & s_{1313} &= \frac{1}{G_{13}}, \quad & s_{1212} &= \frac{1}{G_{12}}, \end{aligned} \tag{4.9}$$

in which Y_1, Y_2 and Y_3 are the Young's moduli associated with three material axes, ν_{12}, ν_{13} and ν_{23} are the Poisson's ratios in the 1-2, 1-3 and 2-3 planes, G_{23}, G_{13} and G_{12} are the shear moduli in the 2-3, 1-3 and 1-2 planes. From the mechanics of material deformation, the Poisson's ratios have the relations

$$\nu_{ij} Y_j = \nu_{ji} Y_i \,. \tag{4.10}$$

Isotropic material can be considered as specific simplification of the orthotropic case, which yields

$$Y = Y_1 = Y_2 = Y_3 \,, \tag{4.11}$$

$$\nu = \nu_{12} = \nu_{13} = \nu_{23} \,, \tag{4.12}$$

$$G = G_{23} = G_{13} = G_{12} = \frac{Y}{2(1+\nu)} \,. \tag{4.13}$$

The third-order tensor of piezoelectric constant tensor $d^{m}_{\cdot kl}$ and the second-order tensor of dielectric constant tensor ϵ^{mn} are arranged as

$$d^{m}_{\cdot kl} = \begin{bmatrix} 0 & 0 & 0 & 0 & d^{1}_{\cdot 13} & 0 \\ 0 & 0 & 0 & d^{2}_{\cdot 23} & 0 & 0 \\ d^{3}_{\cdot 11} & d^{3}_{\cdot 22} & d^{3}_{\cdot 33} & 0 & 0 & 0 \end{bmatrix}, \quad [\epsilon^{mn}] = \begin{bmatrix} \epsilon^{11} & 0 & 0 \\ 0 & \epsilon^{22} & 0 \\ 0 & 0 & \epsilon^{33} \end{bmatrix}. \tag{4.14}$$

In $d^{m}_{\cdot kl}$, the superscript m represent the direction of electric field applied on piezoelectric material, while the subscript pair kl is the stress direction due to the driving electric field.

Alternatively, the constitutive relations of piezoelectric materials can be expressed by stiffness way

$$\sigma^{ij} = c^{ijkl} \varepsilon_{kl} - e^{ijm} E_m \,, \tag{4.15}$$

$$D^{m} = e^{mkl} \varepsilon_{kl} + \chi^{mn} E_n \,, \tag{4.16}$$

where (4.15) is the actuator equation and (4.16) is the sensor equation.

Using the Voigt notations, the components of the fourth-order elasticity constant tensor in (4.15) can be arranged in matrix form as

$$[\breve{c}^{ijkl}] = \begin{bmatrix} \breve{c}^{1111} & \breve{c}^{1122} & \breve{c}^{1133} & 0 & 0 & 0 \\ \breve{c}^{1122} & \breve{c}^{2222} & \breve{c}^{2233} & 0 & 0 & 0 \\ \breve{c}^{1133} & \breve{c}^{2233} & \breve{c}^{3333} & 0 & 0 & 0 \\ 0 & 0 & 0 & \breve{c}^{2323} & 0 & 0 \\ 0 & 0 & 0 & 0 & \breve{c}^{1313} & 0 \\ 0 & 0 & 0 & 0 & 0 & \breve{c}^{1212} \end{bmatrix} \tag{4.17}$$

with

$$\begin{aligned}
\breve{c}^{1111} &= Y_1 \frac{1-\nu_{23}\nu_{32}}{\Delta}, & \breve{c}^{2222} &= Y_2 \frac{1-\nu_{31}\nu_{13}}{\Delta}, \\
\breve{c}^{3333} &= Y_3 \frac{1-\nu_{12}\nu_{21}}{\Delta}, & \breve{c}^{1122} &= Y_1 \frac{\nu_{21}-\nu_{31}\nu_{23}}{\Delta}, \\
\breve{c}^{1133} &= Y_3 \frac{\nu_{13}-\nu_{12}\nu_{23}}{\Delta}, & \breve{c}^{2233} &= Y_2 \frac{\nu_{32}-\nu_{12}\nu_{31}}{\Delta}, \\
\breve{c}^{2323} &= G_{23}, & \breve{c}^{1313} &= G_{13}, \\
\breve{c}^{1212} &= G_{12},
\end{aligned} \tag{4.18}$$

where $\Delta = 1 - \nu_{12}\nu_{21} - \nu_{23}\nu_{32} - \nu_{31}\nu_{13} - \nu_{21}\nu_{32}\nu_{13}$, Y_i denotes the Young's moduli, G_{ij} the shear moduli, and ν_{ij} the Poisson's ratios.

The components of the third-order piezoelectric constant tensor and the second-order dielectric constant tensor in (4.15) and (4.16) can be obtained by

$$e^{mkl} = d^{m}_{\cdot ij} c^{ijkl}, \tag{4.19}$$

$$\chi^{mn} = \epsilon^{mn} - d^{m}_{\cdot ij} e^{ijn}. \tag{4.20}$$

Similarly, they can be arranged by matrix form in

$$[e^{mkl}] = \begin{bmatrix} 0 & 0 & 0 & 0 & e^{113} & 0 \\ 0 & 0 & 0 & e^{223} & 0 & 0 \\ e^{311} & e^{322} & e^{333} & 0 & 0 & 0 \end{bmatrix}, \quad [\chi^{mn}] = \begin{bmatrix} \chi^{11} & 0 & 0 \\ 0 & \chi^{22} & 0 \\ 0 & 0 & \chi^{33} \end{bmatrix}. \tag{4.21}$$

4.3 Coordinate Transformation in Plates and Shells

In the simulation of piezo-laminated plates and shells, isotropic and orthotropic materials are mostly used in the analysis. We define two coordinate systems, one is material coordinate system, denoted by $\breve{\Theta}^i$; the other one is curvilinear coordinate system representing structural geometries, denoted by Θ^i. For isotropic material, the material coordinate axes can be set the same as the curvilinear coordinate axes. However, in case that the fiber reinforcement direction of orthotropic material is not parallel to the curvilinear coordinate axes, like in the case shown in Fig. 4.3, a transformation matrix is necessary for converting the constitutive equations from the material coordinate axes to the curvilinear coordinate axes.

The components of the elasticity constant tensor used in (4.15) are associated with the unit covariant base vectors $\breve{\boldsymbol{i}}_a$ in the material coordinate system. They must be transformed to the base vectors $\boldsymbol{g}_i$, since the formulations of strain field are developed in the curvilinear coordinate system. The transformation matrix is determined by means of the following equations

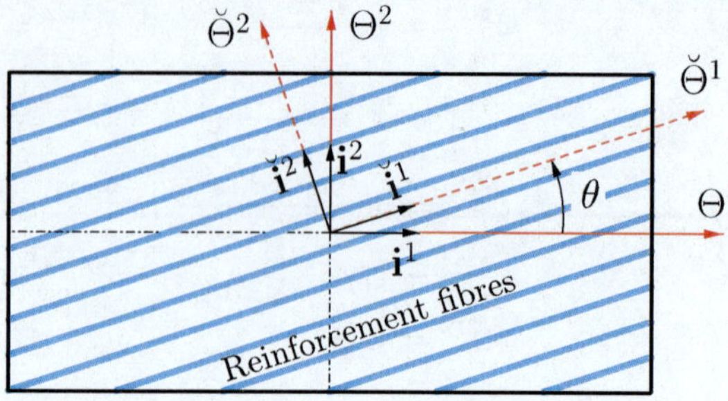

Fig. 4.3 Orientation of reinforcement fibers

$$\begin{aligned} \boldsymbol{c} &= \breve{c}^{abcd}\, \breve{\boldsymbol{i}}_a \otimes \breve{\boldsymbol{i}}_b \otimes \breve{\boldsymbol{i}}_c \otimes \breve{\boldsymbol{i}}_d \\ &= c^{ijkl}\, \boldsymbol{g}_i \otimes \boldsymbol{g}_j \otimes \boldsymbol{g}_k \otimes \boldsymbol{g}_l \,, \end{aligned} \tag{4.22}$$

which leads to

$$c^{ijkl} = \left(\boldsymbol{g}^i \cdot \breve{\boldsymbol{i}}_a\right)\left(\boldsymbol{g}^j \cdot \breve{\boldsymbol{i}}_b\right)\left(\boldsymbol{g}^k \cdot \breve{\boldsymbol{i}}_c\right)\left(\boldsymbol{g}^l \cdot \breve{\boldsymbol{i}}_d\right)\breve{c}^{abcd} \,. \tag{4.23}$$

Here, the indices, a, b, c and d, have the same function as i, j, k and l, but they are used for the components in material coordinate system. Using the same rule one obtains

$$\breve{\varepsilon}_{ab} = \left(\boldsymbol{g}^i \cdot \breve{\boldsymbol{i}}_a\right)\left(\boldsymbol{g}^j \cdot \breve{\boldsymbol{i}}_b\right)\varepsilon_{ij} \,, \tag{4.24}$$

$$\sigma^{ij} = \left(\boldsymbol{g}^i \cdot \breve{\boldsymbol{i}}_a\right)\left(\boldsymbol{g}^j \cdot \breve{\boldsymbol{i}}_b\right)\breve{\sigma}^{ab} \,, \tag{4.25}$$

$$\breve{E}_a = \left(\boldsymbol{g}^i \cdot \breve{\boldsymbol{i}}_a\right) E_i \,, \tag{4.26}$$

$$D^i = \left(\boldsymbol{g}^i \cdot \breve{\boldsymbol{i}}_a\right)\breve{D}^a \,, \tag{4.27}$$

which can be expressed in matrix form as

$$\breve{\boldsymbol{\varepsilon}} = \boldsymbol{T}\boldsymbol{\varepsilon} \,, \quad \boldsymbol{\sigma} = \boldsymbol{T}^{\mathrm{T}}\breve{\boldsymbol{\sigma}} \,, \tag{4.28}$$

$$\breve{\boldsymbol{E}} = \boldsymbol{Q}\boldsymbol{E} \,, \quad \boldsymbol{D} = \boldsymbol{Q}^{\mathrm{T}}\breve{\boldsymbol{D}} \,. \tag{4.29}$$

Due to the neglect of the transverse normal strain $\breve{\varepsilon}_{33}$, the constitutive equations given in (4.15) and (4.16) are simplified to contain only five components, which can be expressed in matrix form as

$$\breve{\boldsymbol{\sigma}} = \breve{\boldsymbol{c}}\breve{\boldsymbol{\varepsilon}} - \breve{\boldsymbol{e}}^{\mathrm{T}}\breve{\boldsymbol{E}} \,, \tag{4.30}$$

$$\breve{\boldsymbol{D}} = \breve{\boldsymbol{e}}\breve{\boldsymbol{\varepsilon}} + \breve{\boldsymbol{\chi}}\breve{\boldsymbol{E}} \,, \tag{4.31}$$

in which the stress vector $\breve{\boldsymbol{\sigma}}$, the strain vector $\breve{\boldsymbol{\varepsilon}}$, the electric displacement vector $\breve{\boldsymbol{D}}$, and the electric field vector $\breve{\boldsymbol{E}}$ are organized as

$$\breve{\boldsymbol{\sigma}} = \begin{Bmatrix} \breve{\sigma}^{11} \\ \breve{\sigma}^{22} \\ \breve{\sigma}^{12} \\ \breve{\sigma}^{23} \\ \breve{\sigma}^{12} \end{Bmatrix} , \quad \breve{\boldsymbol{\varepsilon}} = \begin{Bmatrix} \breve{\varepsilon}_{11} \\ \breve{\varepsilon}_{22} \\ 2\breve{\varepsilon}_{12} \\ 2\breve{\varepsilon}_{23} \\ 2\breve{\varepsilon}_{13} \end{Bmatrix} , \quad \breve{\boldsymbol{D}} = \begin{Bmatrix} \breve{D}^1 \\ \breve{D}^2 \\ \breve{D}^3 \end{Bmatrix} , \quad \breve{\boldsymbol{E}} = \begin{Bmatrix} \breve{E}_1 \\ \breve{E}_2 \\ \breve{E}_3 \end{Bmatrix} . \tag{4.32}$$

In (4.30) and (4.31), $\breve{\boldsymbol{c}}$ denotes the elasticity constant matrix, $\breve{\boldsymbol{d}}$ and $\breve{\boldsymbol{e}}$ are the piezoelectric constant matrices, and $\breve{\boldsymbol{\epsilon}}$ the dielectric constant matrix. The elasticity constant matrix is given by

$$\breve{\boldsymbol{c}} = \begin{bmatrix} \breve{c}_{11} & \breve{c}_{12} & 0 & 0 & 0 \\ \breve{c}_{12} & \breve{c}_{22} & 0 & 0 & 0 \\ 0 & 0 & \breve{c}_{66} & 0 & 0 \\ 0 & 0 & 0 & \breve{c}_{44} & 0 \\ 0 & 0 & 0 & 0 & \breve{c}_{55} \end{bmatrix} , \tag{4.33}$$

with

$$\begin{aligned} \breve{c}_{11} &= \frac{Y_1}{1-\nu_{12}\nu_{21}} , & \breve{c}_{22} &= \frac{Y_2}{1-\nu_{12}\nu_{21}} , & \breve{c}_{12} &= \frac{\nu_{12}Y_2}{1-\nu_{12}\nu_{21}} , \\ \breve{c}_{66} &= G_{12} , & \breve{c}_{55} &= \kappa G_{13} , & \breve{c}_{44} &= \kappa G_{23} . \end{aligned} \tag{4.34}$$

Here, κ is the shear correction factor, which is usually given as $\frac{5}{6}$ or $\frac{\pi}{12}$. The relations between piezoelectric and dielectric constant matrices are

$$\breve{\boldsymbol{e}} = \breve{\boldsymbol{d}}\breve{\boldsymbol{c}} , \tag{4.35}$$

$$\breve{\boldsymbol{\chi}} = \breve{\boldsymbol{\epsilon}} - \breve{\boldsymbol{d}}\breve{\boldsymbol{e}}^{\mathrm{T}} . \tag{4.36}$$

The piezoelectric constant matrix $\breve{\boldsymbol{d}}$ and the dielectric constant matrix $\breve{\boldsymbol{\epsilon}}$ are given

$$\breve{\boldsymbol{d}} = \begin{bmatrix} 0 & 0 & 0 & 0 & \breve{d}_{15} \\ 0 & 0 & 0 & \breve{d}_{24} & 0 \\ \breve{d}_{31} & \breve{d}_{32} & 0 & 0 & 0 \end{bmatrix} , \quad \breve{\boldsymbol{\epsilon}} = \begin{bmatrix} \breve{\epsilon}_{11} & 0 & 0 \\ 0 & \breve{\epsilon}_{22} & 0 \\ 0 & 0 & \breve{\epsilon}_{33} \end{bmatrix} . \tag{4.37}$$

With the help of the transformation matrix given in (4.28) and (4.29), one obtains the constitutive equations described in a curvilinear coordinate system as

$$\boldsymbol{\sigma} = \boldsymbol{c}\boldsymbol{\varepsilon} - \boldsymbol{e}^{\mathrm{T}}\boldsymbol{E} , \tag{4.38}$$

$$\boldsymbol{D} = \boldsymbol{e}\boldsymbol{\varepsilon} + \boldsymbol{\chi}\boldsymbol{E} , \tag{4.39}$$

where

$$\boldsymbol{c} = \boldsymbol{T}^{\mathrm{T}}\breve{\boldsymbol{c}}\boldsymbol{T}\,, \quad \boldsymbol{e} = \boldsymbol{Q}^{\mathrm{T}}\breve{\boldsymbol{e}}\boldsymbol{T}\,, \quad \boldsymbol{\epsilon} = \boldsymbol{Q}^{\mathrm{T}}\breve{\boldsymbol{\epsilon}}\boldsymbol{Q}\,. \tag{4.40}$$

The transformation matrix $\boldsymbol{Q}$ is an identity matrix if the electrical coordinate axes are parallel to the curvilinear coordinate lines, and $\boldsymbol{T}$ is given by

$$\boldsymbol{T} = \begin{bmatrix} t_{11}^2 & t_{21}^2 & t_{11}t_{21} & 0 & 0 \\ t_{12}^2 & t_{22}^2 & t_{12}t_{22} & 0 & 0 \\ 2t_{11}t_{12} & 2t_{21}t_{22} & t_{11}t_{22} + t_{12}t_{21} & 0 & 0 \\ 0 & 0 & 0 & t_{22} & t_{12} \\ 0 & 0 & 0 & t_{21} & t_{11} \end{bmatrix}, \tag{4.41}$$

with

$$\begin{aligned} t_{11} &= \boldsymbol{g}^1 \cdot \breve{\boldsymbol{i}}_1 = \sqrt{g^{11}}\cos\theta\,, & t_{12} &= \boldsymbol{g}^1 \cdot \breve{\boldsymbol{i}}_2 = -\sqrt{g^{11}}\sin\theta\,, \\ t_{21} &= \boldsymbol{g}^2 \cdot \breve{\boldsymbol{i}}_1 = \sqrt{g^{22}}\sin\theta\,, & t_{22} &= \boldsymbol{g}^2 \cdot \breve{\boldsymbol{i}}_2 = \sqrt{g^{22}}\cos\theta\,. \end{aligned} \tag{4.42}$$

4.4 Constitutive Relations for Macro-fiber Composites

4.4.1 Configurations of Macro-fiber Composites

Macro-fiber composites mainly consist of piezoceramic fibers, epoxy matrix and electrodes. They have two different modes of structures, namely d31 or d33 modes. The first mode of MFC material is abbreviated as MFC-d31. The piezoelectric fiber is oriented in the in-plane direction, and the polarization is pointing along the thickness direction. Thus in the first mode the d_{31} effect is dominating the actuation forces. The second type of MFC material is denoted as MFC-d33. MFC-d33 is arranged in a specific manner such that the polarization of the piezoelectric material is along the piezo-fiber direction. Therefore, MFC-d33 mainly uses the d_{33} effect for generation of actuation forces. Because the coefficient d_{33} is usually much larger (about 2 times larger) than d_{31}. MFC-d33 patches have larger actuation forces than MFC-d31 ones. Additionally, actuation voltages for MFC-d31 patches can be applied in the range from only −60 to 360 V (with the electrode distance of 0.18 mm), while those for MFC-d33 patches can vary between −500 and 1500 V (with center-to-center interdigitated electrode spacing of 0.5 mm) [10]. The schematic of these two kinds of MFCs are shown in Fig. 4.4a and b, respectively.

The special arrangement of MFC material increases the structural flexibility. The interdigitated electrodes reduce the impact on structural performance due to damage or fracture of the piezoceramics or electrode. Any damage on piezoceramics or electrode will not influence significantly on the overall actuation effect.

In order to present clearly the modeling procedure, three coordinate systems are defined, as can be seen in Fig. 4.4, namely the curvilinear coordinate system repre-

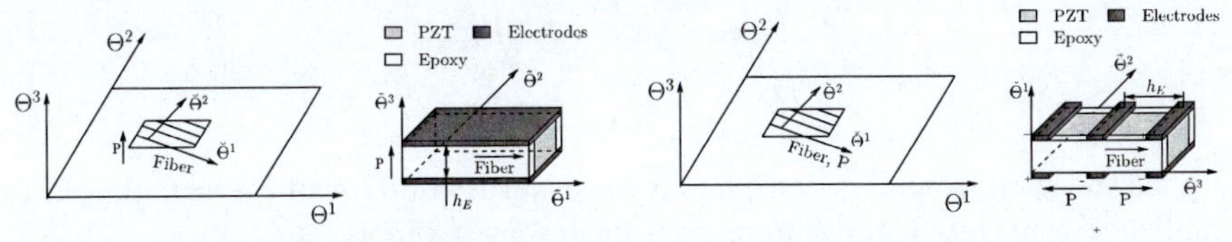

(a) MFC-d31 structure (b) MFC-d33 structure

Fig. 4.4 Schematic of different kinds of MFC models

sented by Θ^i $(i = 1, 2, 3)$, the material coordinate system (also called fiber coordinate system) denoted by $\breve{\Theta}^i$, and the polarization coordinate system shown as $\tilde{\Theta}^i$.

The curvilinear coordinate system is usually used for representing the geometry of thin-walled structures, in which the thickness direction is defined as the Θ^3-line, the Θ^1- and Θ^2-line defines the in-plane directions. The fiber coordinate system defines the fiber orientation in both MFC and composite materials. The $\breve{\Theta}^1$-line defines the fiber alignment; the $\breve{\Theta}^2$-line is normal to the fiber alignment in the in-plane dimension; the $\breve{\Theta}^3$-line is along the thickness direction. The angle between Θ^1 and $\breve{\Theta}^1$ defines the fiber angle, which is a parameter in the transformation matrix. The polarization coordinate system is used for MFC material, in which the $\tilde{\Theta}^3$-line is pointing along the direction of polarization of piezoelectric material.

MFC materials are usually appeared in the form of layers or patches. Even though the composition and structural arrangement of MFC materials are very complex, they can be homogenized to an orthotropic material model, see e.g. [11–15] among others. For more details of structural design of MFC material, we refer to [4–6, 10].

4.4.2 Constitutive for Plates and Shells

Considering small strains and weak electric field in piezoelectric patches or layers, the linear constitutive equations coupled with electric and mechanical fields can be expressed in the fiber coordinate system as [9]

$$\breve{\varepsilon}_{ij} = \breve{s}_{ijkl}\breve{\sigma}_{kl} + \breve{d}_{ijm}\breve{E}_m \,, \tag{4.43}$$

$$\breve{D}_m = \breve{d}_{mkl}\breve{\sigma}_{kl} + \breve{\epsilon}_{mn}\breve{E}_n \,. \tag{4.44}$$

Here $\breve{\varepsilon}_{ij}, \breve{\sigma}_{kl}, \breve{D}_m, \breve{E}_m, \breve{s}_{ijkl}, \breve{d}_{mkl}$ and $\breve{E}_n$ are measured in the fibrous coordinate system, which have the same meaning as those introduced in Sect. 4.2. For simplicity all the indices are in the lower position.

Using the Voigt notations, as shown in Table 4.1, Eqs. (4.43) and (4.44) can be written as

$$\breve{\varepsilon}_p = \breve{s}_{pq}\breve{\sigma}_q + \breve{d}_{pm}\breve{E}_m, \tag{4.45}$$

$$\breve{D}_m = \breve{d}_{mq}\breve{\sigma}_q + \breve{\epsilon}_{mn}\breve{E}_n. \tag{4.46}$$

For plate and shell structures, introducing the usual assumption of $\breve{\sigma}_{33} = 0$, the elastic compliance constants $\breve{s}_{pq}$ in fibrous coordinates are given as

$$\begin{aligned} &\breve{s}_{11} = \frac{1}{\breve{Y}_1}, \quad \breve{s}_{12} = -\frac{\breve{\nu}_{12}}{\breve{Y}_1} = -\frac{\breve{\nu}_{21}}{\breve{Y}_2}, \quad \breve{s}_{22} = \frac{1}{\breve{Y}_2}, \\ &\breve{s}_{44} = \frac{1}{\kappa\breve{G}_{23}}, \quad \breve{s}_{55} = \frac{1}{\kappa\breve{G}_{13}}, \quad \breve{s}_{66} = \frac{1}{\breve{G}_{12}}, \end{aligned} \tag{4.47}$$

where $\breve{Y}_i$, $\breve{\nu}_{ij}$ and $\breve{G}_{ij}$ are the Young's moduli, the Poisson's ratios and the shear moduli, κ is the shear correction factor.

Re-arranging the constitutive equations (4.45) and (4.46) by the matrix form with an inversed relation, one obtains

$$\breve{\boldsymbol{\sigma}} = \breve{\boldsymbol{c}}\breve{\boldsymbol{\varepsilon}} - \breve{\boldsymbol{e}}^{\mathrm{T}}\breve{\boldsymbol{E}}, \tag{4.48}$$

$$\breve{\boldsymbol{D}} = \breve{\boldsymbol{e}}\breve{\boldsymbol{\varepsilon}} + \breve{\boldsymbol{\chi}}\breve{\boldsymbol{E}}, \tag{4.49}$$

where

$$\breve{\boldsymbol{\sigma}} = \begin{Bmatrix} \breve{\sigma}_{11} \\ \breve{\sigma}_{22} \\ \breve{\tau}_{12} \\ \breve{\tau}_{23} \\ \breve{\tau}_{13} \end{Bmatrix}, \quad \breve{\boldsymbol{\varepsilon}} = \begin{Bmatrix} \breve{\varepsilon}_{11} \\ \breve{\varepsilon}_{22} \\ \breve{\gamma}_{12} \\ \breve{\gamma}_{23} \\ \breve{\gamma}_{13} \end{Bmatrix}, \quad \breve{\boldsymbol{D}} = \begin{Bmatrix} \breve{D}_1 \\ \breve{D}_2 \\ \breve{D}_3 \end{Bmatrix}, \quad \breve{\boldsymbol{E}} = \begin{Bmatrix} \breve{E}_1 \\ \breve{E}_2 \\ \breve{E}_3 \end{Bmatrix}, \tag{4.50}$$

$$\breve{\boldsymbol{c}} = \begin{bmatrix} \breve{c}_{11} & \breve{c}_{12} & 0 & 0 & 0 \\ \breve{c}_{12} & \breve{c}_{22} & 0 & 0 & 0 \\ 0 & 0 & \breve{c}_{66} & 0 & 0 \\ 0 & 0 & 0 & \breve{c}_{44} & 0 \\ 0 & 0 & 0 & 0 & \breve{c}_{55} \end{bmatrix}, \quad \breve{\boldsymbol{\chi}} = \begin{bmatrix} \breve{\chi}_{11} & 0 & 0 \\ 0 & \breve{\chi}_{22} & 0 \\ 0 & 0 & \breve{\chi}_{33} \end{bmatrix}, \tag{4.51}$$

with

$$\begin{aligned} &\breve{c}_{11} = \frac{\breve{s}_{22}}{\breve{s}_{11}\breve{s}_{22} - \breve{s}_{12}\breve{s}_{12}} = \frac{\breve{Y}_1}{1 - \breve{\nu}_{12}\breve{\nu}_{21}}, \\ &\breve{c}_{12} = -\frac{\breve{s}_{12}}{\breve{s}_{11}\breve{s}_{22} - \breve{s}_{12}\breve{s}_{12}} = \frac{\breve{\nu}_{12}\breve{Y}_2}{1 - \breve{\nu}_{12}\breve{\nu}_{21}}, \\ &\breve{c}_{22} = \frac{\breve{s}_{11}}{\breve{s}_{11}\breve{s}_{22} - \breve{s}_{12}\breve{s}_{12}} = \frac{\breve{Y}_2}{1 - \breve{\nu}_{12}\breve{\nu}_{21}}, \\ &\breve{c}_{44} = \kappa\breve{G}_{23}, \quad \breve{c}_{55} = \kappa\breve{G}_{13}, \quad \breve{c}_{66} = \breve{G}_{12}. \end{aligned} \tag{4.52}$$

Here $\breve{\boldsymbol{\sigma}}$, $\breve{\boldsymbol{\varepsilon}}$ denote the stress and strain vectors, $\breve{\boldsymbol{c}}$ is the elasticity constant matrix. Furthermore, $\breve{\boldsymbol{D}}$, $\breve{\boldsymbol{E}}$, $\breve{\boldsymbol{e}}$ and $\breve{\boldsymbol{\chi}}$ represent the electric displacement vector, the electric field vector, the piezoelectric constant matrix and the dielectric constant matrix, respectively, among which $\breve{\boldsymbol{e}}$ depends strongly on the structure of MFC materials.

4.4.3 *Piezo Constants for MFC-d31 Type*

Regarding to MFC-d31 material, with the structural arrangement shown in Fig. 4.4, the polarization is pointing along the thickness direction. In addition, the piezoelectric fiber reinforcement is aligned in the in-plane direction. Therefore, in this case the polarization coordinates are the same with the fiber coordinates. Thus, the piezoelectric constant matrix are organized as

$$\breve{\boldsymbol{e}} = \begin{bmatrix} 0 & 0 & 0 & 0 & \breve{e}_{15} \\ 0 & 0 & 0 & \breve{e}_{24} & 0 \\ \breve{e}_{31} & \breve{e}_{32} & 0 & 0 & 0 \end{bmatrix} . \tag{4.53}$$

MFC materials are usually produced in layers or patches, the electrodes exist only in the outer surfaces, which are parallel to the mid-surface. This means that the electric field can be applied only in the thickness direction. For simplicity, the constitutive equation for the direct effect reduces to one dimension as

$$\breve{D}_3 = \left[\breve{e}_{31}\ \breve{e}_{32}\ 0\ 0\ 0\right] \breve{\boldsymbol{\varepsilon}} + \breve{\chi}_{33} \breve{E}_3 \ , \tag{4.54}$$

with

$$\breve{e}_{31} = \frac{\breve{d}_{31}\breve{s}_{22} - \breve{d}_{32}\breve{s}_{12}}{\breve{s}_{11}\breve{s}_{22} - \breve{s}_{12}\breve{s}_{12}} = \breve{d}_{31}\breve{c}_{11} + \breve{d}_{32}\breve{c}_{12}, \tag{4.55}$$

$$\breve{e}_{32} = \frac{\breve{d}_{31}\breve{s}_{12} - \breve{d}_{32}\breve{s}_{11}}{\breve{s}_{12}\breve{s}_{12} - \breve{s}_{11}\breve{s}_{22}} = \breve{d}_{31}\breve{c}_{12} + \breve{d}_{32}\breve{c}_{22}, \tag{4.56}$$

$$\breve{\chi}_{33} = \breve{\epsilon}_{33} - \breve{d}_{31}\breve{e}_{31} - \breve{d}_{32}\breve{e}_{32}, \tag{4.57}$$

Assuming that the electric potential is linearly distributed through the thickness direction yields constant electric field through the thickness, with the definition of electric field as

$$\breve{E}_3 = -\frac{\breve{\Phi}_3}{h_E}, \tag{4.58}$$

where h_E denotes the distance between two electrodes, and $\breve{\Phi}_3$ is the electric voltage applied along the thickness direction, as shown in Fig. 4.4a.

4.4.4 Piezo Constants for MFC-d33 Type

Compared to MFC-d31 material, even though MFC-d33 material has similar arrangement, but it has different polarization direction in piezoelectric fiber. The polarization direction of MFC-d33 is pointing along the piezoelectric fiber reinforcement. Thus the driving electric field must align in the same or opposite direction of polarization. The piezoelectric constant matrix for MFC-d33 will be organized as

$$\breve{\boldsymbol{e}} = \begin{bmatrix} \breve{e}_{11} & \breve{e}_{12} & 0 & 0 & 0 \\ 0 & 0 & \breve{e}_{26} & 0 & 0 \\ 0 & 0 & 0 & 0 & \breve{e}_{35} \end{bmatrix} . \tag{4.59}$$

In MFC-d33 material, both the polarization and electric field directions align with the piezo fiber orientation. This will lead to MFC-d33 mainly using d33 effect. Because only one pair of electrodes existing in MFC-d33 patches, the electric field can be applied only in the polarization direction. Therefore, in a similar way, the constitutive equation for the direct effect reduces to

$$\breve{D}_1 = \left[\breve{e}_{11}\ \breve{e}_{12}\ 0\ 0\ 0\right] \breve{\boldsymbol{\varepsilon}} + \breve{\chi}_{11} \breve{E}_1 \ , \tag{4.60}$$

with

$$\breve{e}_{11} = \frac{\breve{d}_{11}\breve{s}_{22} - \breve{d}_{12}\breve{s}_{12}}{\breve{s}_{11}\breve{s}_{22} - \breve{s}_{12}\breve{s}_{12}} = \breve{d}_{11}\breve{c}_{11} + \breve{d}_{12}\breve{c}_{12}, \tag{4.61}$$

$$\breve{e}_{12} = \frac{\breve{d}_{11}\breve{s}_{12} - \breve{d}_{12}\breve{s}_{11}}{\breve{s}_{12}\breve{s}_{12} - \breve{s}_{11}\breve{s}_{22}} = \breve{d}_{11}\breve{c}_{12} + \breve{d}_{12}\breve{c}_{22}, \tag{4.62}$$

$$\breve{\chi}_{11} = \breve{\epsilon}_{11} - \breve{d}_{11}\breve{e}_{11} - \breve{d}_{12}\breve{e}_{12}, \tag{4.63}$$

The mode of interdigitated electrodes in MFC-d33 patches are very different with that in MFC-d31. In this arrangement, the electric field is very complex and distributed non-uniformly, in which a certain volume of piezoelectric fiber is inactive. The real electric field distribution along the piezoelectric fiber was deeply investigated by Bowen et al. [16]. For simplicity, the paper follows the work of Williams [17] that the electric field is assumed to be uniform and constant between two electrodes and distributed perfectly through the material, which yields

$$\breve{E}_1 = -\frac{\breve{\Phi}_1}{h_E}. \tag{4.64}$$

Here h_E denotes the distance between two electrodes, which is not equal to the thickness of the MFC-d33 layer, as can be seen in Fig. 4.4b, and $\breve{\Phi}_1$ is the electric voltage applied along the $\breve{\Theta}^1$-axis.

Table 4.2 Description of material parameters for MFC, reprinted from Ref. [18], copyright 2015, with permission from ELSEVIER

MFC in fibrous axes	$\breve{Y}_1$	$\breve{Y}_2$	$\breve{\nu}_{12}$	$\breve{\nu}_{23}$	$\breve{G}_{12}$	$\breve{G}_{13}$	$\breve{G}_{23}$	$\breve{d}_{31}$	$\breve{d}_{32}$	$\breve{\epsilon}_{33}$	$\breve{d}_{11}$	$\breve{d}_{12}$	$\breve{\epsilon}_{11}$
MFC-d31	$\tilde{Y}_1$	$\tilde{Y}_2$	$\tilde{\nu}_{12}$	$\tilde{\nu}_{23}$	$\tilde{G}_{12}$	$\tilde{G}_{13}$	$\tilde{G}_{23}$	$\tilde{d}_{31}$	$\tilde{d}_{32}$	$\tilde{\epsilon}_{33}$	/	/	/
MFC-d33	$\tilde{Y}_3$	$\tilde{Y}_2$	$\tilde{\nu}_{32}$	$\tilde{\nu}_{21}$	$\tilde{G}_{32}$	$\tilde{G}_{31}$	$\tilde{G}_{21}$	/	/	/	$\tilde{d}_{33}$	$\tilde{d}_{32}$	$\tilde{\epsilon}_{11}$

4.4.5 Parameter Configuration

The two modes of MFC materials consist of active layer, electrode layer, protection layer. Each layer can be homogenized to an orthotropic material layer. Using the lamination theory of layered structures, the overall MFC patches can be modeled as orthotropic material. The fiber reinforced direction usually has a larger Young's modulus than the other two directions, and the parameters in the directions normal to the fiber reinforcement are assumed to be equal. Therefore, the equivalent MFC material has 7 elastic material parameters and 3 electrical material parameters, as shown in Table 4.2.

4.4.6 Multi-layer Piezo Composites

Considering multi-layers of MFC materials embedded into laminated structures, as shown in Fig. 4.5, the constitutive equations must be transformed from the fiber coordinate system to the curvilinear coordinate system. Finally, the constitutive equations can be expressed as

$$\boldsymbol{\sigma} = \boldsymbol{c}\boldsymbol{\varepsilon} - \boldsymbol{e}^{\mathrm{T}}\boldsymbol{E}, \tag{4.65}$$

$$\boldsymbol{D} = \boldsymbol{e}\boldsymbol{\varepsilon} + \boldsymbol{\chi}\boldsymbol{E}, \tag{4.66}$$

with

$$\boldsymbol{c} = \boldsymbol{T}^{\mathrm{T}}\breve{\boldsymbol{c}}\boldsymbol{T}, \quad \boldsymbol{e} = \breve{\boldsymbol{e}}\boldsymbol{T}, \quad \boldsymbol{\chi} = \breve{\boldsymbol{\chi}}, \tag{4.67}$$

where $\boldsymbol{T}$ is a transformation matrix, given in Eq. (4.41). Since the electric field is always pointing along the polarization direction, the angle between electric field and polarization is zero, which yields $\boldsymbol{E} = \breve{\boldsymbol{E}}$.

Assuming smart structures with N layers of MFC patches, the vectors $\boldsymbol{D}$, $\boldsymbol{E}$ and the matrices $\boldsymbol{e}$, $\boldsymbol{\chi}$ can be arranged as follows:

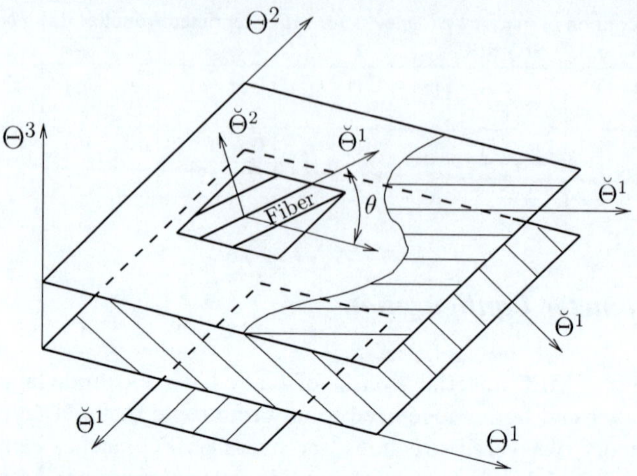

Fig. 4.5 Multi-layer composites with MFCs, reprinted from Ref. [18], copyright 2015, with permission from ELSEVIER

$$\boldsymbol{D} = \begin{Bmatrix} D_s^{(1)} \\ D_s^{(2)} \\ \vdots \\ D_s^{(N)} \end{Bmatrix}, \quad \boldsymbol{E} = \begin{Bmatrix} E_s^{(1)} \\ E_s^{(2)} \\ \vdots \\ E_s^{(N)} \end{Bmatrix}, \tag{4.68}$$

$$\boldsymbol{e} = \begin{bmatrix} e_{s1}^{(1)} & e_{s2}^{(1)} & 0 & 0 & 0 \\ e_{s1}^{(2)} & e_{s2}^{(2)} & 0 & 0 & 0 \\ \vdots & \vdots & \vdots & \vdots & \vdots \\ e_{s1}^{(N)} & e_{s2}^{(N)} & 0 & 0 & 0 \end{bmatrix}, \quad \boldsymbol{\chi} = \begin{bmatrix} \chi_{ss}^{(1)} & 0 & \cdots & 0 \\ 0 & \chi_{ss}^{(2)} & \cdots & 0 \\ \vdots & \vdots & \ddots & \vdots \\ 0 & 0 & \cdots & \chi_{ss}^{(N)} \end{bmatrix}. \tag{4.69}$$

Here the subscript $s = 3$ is for MFC-d31 materials, $s = 1$ is for MFC-d33 materials, and N denotes the total number of MFC layers.

The driving electric field for MFC-d31 is along the thickness direction, and that for MFC-d33 is along the fiber reinforcement direction. In both of these two cases, the electric field in the structural coordinates is the same as that in fiber coordinates. This results in identity transformation matrix for the electric constant matrix from the structural coordinates to the fiber coordinates. Therefore, the electric field vector for multi-layer MFC structures are

$$\boldsymbol{E} = \begin{Bmatrix} E_s^{(1)} \\ E_s^{(2)} \\ \vdots \\ E_s^{(N)} \end{Bmatrix} = \begin{bmatrix} -\frac{1}{h_E^{(1)}} & 0 & \cdots & 0 \\ 0 & -\frac{1}{h_E^{(2)}} & \cdots & 0 \\ \vdots & \vdots & \ddots & \vdots \\ 0 & 0 & \cdots & -\frac{1}{h_E^{(N)}} \end{bmatrix} \begin{Bmatrix} \Phi_s^{(1)} \\ \Phi_s^{(2)} \\ \vdots \\ \Phi_s^{(N)} \end{Bmatrix} = \boldsymbol{B}_\phi \boldsymbol{\Phi}, \tag{4.70}$$

where $\boldsymbol{B}_\phi$ denote the electric field matrix, and $\boldsymbol{\Phi}$ is the electric voltage vector applied on MFC patches.

4.5 Electroelastic Nonlinear Constitutive Relations

Concerning structures deforming in elastic range and under strong electric field, the nonlinear constitutive equations including second-order of electroelastic terms are adopted [19]

$$\varepsilon_{ij} = s_{ijkl}\sigma_{kl} + d_{ijm}E_m + \frac{1}{2}\beta_{ijmn}E_mE_n\,, \tag{4.71}$$

$$D_m = d_{mkl}\sigma_{kl} + \epsilon_{mn}E_n + \frac{1}{2}\chi_{mkn}E_kE_n\,. \tag{4.72}$$

Here, the Latin indices, i, j, k, l, m, n, take the numbers 1, 2 or 3, while ij or kl denote only 11, 22, 33, 12 or 21, 13 or 31, 23 or 32. In (4.71) and (4.72), ε_{ij} and σ_{kl}, denote respectively the strain and stress components, D_m and E_n are the electric displacement and electric field components. The coefficients s_{ijkl}, d_{mkl} and ϵ_{mn} represent, respectively, the tensors of elastic compliance constants, piezoelectric constants and dielectric constants, β_{ijmn} and χ_{mkn} are the nonlinear electroelastic constants and nonlinear electroelastic susceptibility constants, respectively.

Again using the Voigt notation, given in Table 4.1, Eqs. (4.71) and (4.72) can be re-written as

$$\varepsilon_p = s_{pq}\sigma_q + d_{pm}E_m + \frac{1}{2}\beta_{pmn}E_mE_n\,, \tag{4.73}$$

$$D_m = d_{mq}\sigma_q + \epsilon_{mn}E_n + \frac{1}{2}\chi_{mkn}E_kE_n\,. \tag{4.74}$$

Here, the elastic compliance constants s_{pq} are calculated by the material elastic properties as

$$\begin{aligned} s_{11} &= \frac{1}{Y_1}, \quad s_{12} = -\frac{\nu_{12}}{Y_1} = -\frac{\nu_{21}}{Y_2}, \quad s_{22} = \frac{1}{Y_2},\\ s_{44} &= \frac{1}{\kappa G_{23}}, \quad s_{55} = \frac{1}{\kappa G_{13}}, \quad s_{66} = \frac{1}{G_{12}}, \end{aligned} \tag{4.75}$$

where Y_i, ν_{12} and G_{ij} are the Young's moduli, the Poisson's ratios and the shear moduli, and $\kappa = 5/6$ is the shear correction factor.

Assuming each piezoelectric patch has only one pair of electrodes, electric field can be applied in one polarization direction. If the polarization aligns along the thickness direction, it leads to $m = n = k = 3$. Again, because of the characteristic of plates and shells, the transverse normal strain assumes zero, $\sigma_3 = 0$. Therefore, for each layer, the constitutive equations are simplified as

$$\sigma_p = c_{pq}\varepsilon_q - e_{pm}E_m - \frac{1}{2}b_{pmn}E_m E_n\,, \tag{4.76}$$

$$D_m = e_{mq}\varepsilon_q + g_{mn}E_n + \frac{1}{2}h_{mkn}E_k E_n\,. \tag{4.77}$$

with

$$c_{11} = \frac{s_{22}}{s_{11}s_{22} - s_{12}s_{21}} = \frac{Y_1}{1-\nu_{12}\nu_{21}}\,, \tag{4.78}$$

$$c_{12} = -\frac{s_{12}}{s_{11}s_{22} - s_{12}s_{21}} = \frac{\nu_{12}Y_2}{1-\nu_{12}\nu_{21}}\,, \tag{4.79}$$

$$c_{22} = \frac{s_{11}}{s_{11}s_{22} - s_{12}s_{21}} = \frac{Y_2}{1-\nu_{12}\nu_{21}}\,, \tag{4.80}$$

$$c_{44} = \kappa G_{23}, \quad c_{55} = \kappa G_{13}, \quad c_{66} = G_{12}, \tag{4.81}$$

$$e_{31} = \frac{d_{31}s_{22} - d_{32}s_{12}}{s_{11}s_{22} - s_{12}s_{21}} = d_{31}c_{11} + d_{32}c_{12}\,, \tag{4.82}$$

$$e_{32} = \frac{d_{31}s_{21} - d_{32}s_{11}}{s_{12}s_{21} - s_{11}s_{22}} = d_{31}c_{21} + d_{32}c_{22}\,, \tag{4.83}$$

$$b_{331} = \frac{\beta_{331}s_{22} - \beta_{332}s_{12}}{s_{11}s_{22} - s_{12}s_{21}} = \beta_{331}c_{11} + \beta_{332}c_{12}\,, \tag{4.84}$$

$$b_{332} = \frac{\beta_{331}s_{21} - \beta_{332}s_{11}}{s_{12}s_{21} - s_{11}s_{22}} = \beta_{331}c_{21} + \beta_{332}c_{22}\,, \tag{4.85}$$

$$g_{33} = \epsilon_{33} - d_{31}e_{31} - d_{32}e_{32}\,, \tag{4.86}$$

$$h_{333} = \chi_{333} - d_{31}b_{331} - d_{32}b_{332}\,. \tag{4.87}$$

In the case of multi-layer structures with piezoelectric patches and cross- or angle-ply laminated composites, as shown in Fig. 4.5, the constitutive equations must transform from the fiber coordinate system to the curvilinear coordinate system by transformation matrix $\boldsymbol{T}$, given in (4.41). Thus the constitutive equations in matrix form referred to the curvilinear coordinate system are

$$\boldsymbol{\sigma} = \boldsymbol{c}\boldsymbol{\varepsilon} - \boldsymbol{e}^{\mathrm{T}}\boldsymbol{E} - \frac{1}{2}\boldsymbol{b}|\bar{\boldsymbol{E}}|\boldsymbol{E}, \tag{4.88}$$

$$\boldsymbol{D} = \boldsymbol{e}\boldsymbol{\varepsilon} + \boldsymbol{g}\boldsymbol{E} + \frac{1}{2}\boldsymbol{h}|\bar{\boldsymbol{E}}|\boldsymbol{E}. \tag{4.89}$$

Here

$$\boldsymbol{\sigma} = \begin{Bmatrix} \sigma_{11} \\ \sigma_{22} \\ \tau_{12} \\ \tau_{23} \\ \tau_{13} \end{Bmatrix}, \quad \boldsymbol{\varepsilon} = \begin{Bmatrix} \varepsilon_{11} \\ \varepsilon_{22} \\ \gamma_{12} \\ \gamma_{23} \\ \gamma_{13} \end{Bmatrix}, \quad \boldsymbol{D} = \begin{Bmatrix} D_3^{(1)} \\ D_3^{(2)} \\ \vdots \\ D_3^{(N)} \end{Bmatrix}, \quad \boldsymbol{E} = \begin{Bmatrix} E_3^{(1)} \\ E_3^{(2)} \\ \vdots \\ E_3^{(N)} \end{Bmatrix}, \tag{4.90}$$

$$\boldsymbol{c} = \begin{bmatrix} c_{11} & c_{12} & 0 & 0 & 0 \\ c_{12} & c_{22} & 0 & 0 & 0 \\ 0 & 0 & c_{66} & 0 & 0 \\ 0 & 0 & 0 & c_{44} & 0 \\ 0 & 0 & 0 & 0 & c_{55} \end{bmatrix}, \quad \bar{\boldsymbol{E}} = \begin{bmatrix} E_3^{(1)} & 0 & \cdots & 0 \\ 0 & E_3^{(2)} & \cdots & 0 \\ \vdots & \vdots & \ddots & \vdots \\ 0 & 0 & \cdots & E_3^{(N)} \end{bmatrix}, \tag{4.91}$$

$$\boldsymbol{e} = \begin{bmatrix} e_{31}^{(1)} & e_{32}^{(1)} & 0 & 0 & 0 \\ e_{31}^{(2)} & e_{32}^{(2)} & 0 & 0 & 0 \\ \vdots & \vdots & \vdots & \vdots & \vdots \\ e_{31}^{(N)} & e_{32}^{(N)} & 0 & 0 & 0 \end{bmatrix}, \quad \boldsymbol{b} = \begin{bmatrix} b_{331}^{(1)} & b_{332}^{(1)} & 0 & 0 & 0 \\ b_{331}^{(2)} & b_{332}^{(2)} & 0 & 0 & 0 \\ \vdots & \vdots & \vdots & \vdots & \vdots \\ b_{331}^{(N)} & b_{332}^{(N)} & 0 & 0 & 0 \end{bmatrix}, \tag{4.92}$$

$$\boldsymbol{g} = \begin{bmatrix} g_{33}^{(1)} & 0 & \cdots & 0 \\ 0 & g_{33}^{(2)} & \cdots & 0 \\ \vdots & \vdots & \ddots & \vdots \\ 0 & 0 & \cdots & g_{33}^{(N)} \end{bmatrix}, \quad \boldsymbol{h} = \begin{bmatrix} h_{333}^{(1)} & 0 & \cdots & 0 \\ 0 & h_{333}^{(2)} & \cdots & 0 \\ \vdots & \vdots & \ddots & \vdots \\ 0 & 0 & \cdots & h_{333}^{(N)} \end{bmatrix}, \tag{4.93}$$

In the above equations, N represents the total number of piezoelectric layers, $\boldsymbol{\sigma}$, $\boldsymbol{\varepsilon}$, $\boldsymbol{D}$, $\boldsymbol{E}$ are the stress vector, the strain vector, the electric displacement vector, the electric field vector; $\boldsymbol{c}$ is the elasticity constant matrix; $\bar{\boldsymbol{E}}$ is the nonlinear electric field coefficient matrix; $\boldsymbol{e}$ and $\boldsymbol{g}$ denote, respectively, the piezoelectric constant and dielectric constant matrices; $\boldsymbol{b}$ and $\boldsymbol{h}$ are the nonlinear electroelastic strain constant and susceptibility constant matrices.

Considering structures undergoing large displacements but in elastic range and under strong electric field, both geometrically nonlinear and electro-elastic nonlinear effects should be taken into account. The resulting nonlinear models are abbreviated as RVK5SE, MRT5SE, LER5SE, LRT56SE, where SE represents strong electric field. The model including geometrically linear and electro-elastic nonlinear phenomena is denoted by LIN5SE. If linear constitutive equations are considered, the resulting models are denoted by LIN5WE, RVK5WE, MRT5WE, LER5WE, LRT56WE, in which WE is shortened by weak electric filed. In most of this report, the WE is not always appearing in the model abbreviations. If the model abbreviations exclude WE, then the model considers only linear constitutive equations.

4.6 Summary

This chapter dealt with constitutive equations for both isotropic or orthotropic piezoelectric materials. Piezoelectricity was discussed for the deep insight of piezoelectric materials, which is followed by fundamental theory of piezoelectricity in three-dimensional space. Later, the constitutive equations for plates and shells were constructed, with the transformation law between the fiber coordinates and structural coordinates. Two typical constitutive equations of MFC materials were developed for multi-layered MFC structures. Finally, an electroelastic coupled nonlinear constitutive relations was presented for the simulation of structures under strong electric filed.

References

1. H. Kawai, The piezoelectricity of poly(vinylidene) fluoride. Jpn. J. Appl. Phys. **8**, 975–976 (1969)
2. R.B. Williams, B.W. Grimsley, D.J. Inman, W.K. Wilkie, Manufacturing and mechanics-based characterization of macro fiber composite actuators, in *ASME 2002 International Mechanical Engineering Congress and Exposition* (2002), pp. 79–89
3. W.K. Wilkie, R.G. Bryant, J.W. High, R.L. Fox, R.F. Hellbaum, A. Jalink, B.D. Little, P.H. Mirick, Low-cost piezocomposite actuator for structural control applications, in *SPIE - Smart Structures and Materials 2000: Industrial and Commercial Applications of Smart Structures Technologies*, vol. 3991 (SPIE, 12 June 2000), pp. 323–334
4. R.B. Williams, W.K. Wilkie, D.J. Inman, An overview of composite actuators with piezoceramic fibers, in *Proceedings of IMAC-XX: Conference & Exposition on Structural Dynamics*, vol. 4753 (Los Angeles, CA; United States, 4-7 February 2002), pp. 421–427
5. H.A. Sodano, J. Lloyd, D.J. Inman, An experimental comparison between several active composite actuators for power generation. Smart Mater. Struct. **15**, 1211–1216 (2006)
6. C.R. Bowen, R. Stevens, L.J. Nelson, A.C. Dent, G. Dolman, B. Su, T.W. Button M.G. Cain, M. Stewart, Manufacture and characterization of high activity piezoelectric fibres. Smart Mater. Struct. **15**, 295–301 (2006)
7. P. Gaudenzi, *Smart Structures: Physical Behavior, Mathematical Modeling and Applications* (A John Wiley & Sons Ltd., Publication, 2009)
8. V. Piefort, *Finite element modeling of piezoelectric active structures*. Ph.D. Thesis, Universite Libre de Bruxelles (2001)
9. H.F. Tiersten, Electroelastic equations for electroded thin plates subject to large driving voltages. J. Appl. Phys. **74**, 3389–3393 (1993)
10. Smart Material Corp. www.smart-material.com
11. A. Deraemaeker, S. Benelechi, A. Benjeddou, A. Preumont, Analytical and numerical computation of homogenized properties of MFCs: application to a composite boom with MFC actuators and sensors, in *Proceedings of the III ECCOMAS Thematic Conference on Smart Structures and Materials* (Gdansk, Poland, 9-11 July 2007)
12. A. Deraemaeker, H. Nasser, A. Benjeddou, A. Preumont, Mixing rules for the piezoelectric properties of macro fiber composites. J. Intell. Mater. Syst. Struct. **20**(12), 1475–1482 (2009)
13. A. Deraemaeker, H. Nasser, Numerical evaluation of the equivalent properties of macro fiber composite (MFC) transducers using periodic homogenization. Int. J. Solids Struct. **47**, 3272–3285 (2010)

14. F. Biscani, H. Nasser, S. Belouettar, E. Carrera, Equivalent electro-elastic properties of macro fiber composite (MFC) transducers using asymptotic expansion approach. Compos. Part B **42**, 444–455 (2011)
15. Y.X. Li, S.Q. Zhang, R. Schmidt, X.S. Qin, Homogenization for macro-fiber composites using Reissner-Mindlin plate theory, in *Journal of Intelligent Material Systems and Structures* (2016)
16. C.R. Bowen, L.J. Nelson, R. Stevens, M.G. Cain, M. Stewart, Optimisation of interdigitated electrodes for piezoelectric actuators and active fibre composites. J. Electroceramics **16**(4), 263–269 (2006)
17. R.B. Williams, *Nonlinear mechanical and actuation characterization of piezoceramic fiber composites*. PhD thesis, Virginia Polytechnic Institute and State University (2004)
18. S.Q. Zhang, Y.X. Li, R. Schmidt, Modeling and simulation of macro-fiber composite layered smart structures. Compos. Struct. **126**, 89–100 (2015)
19. H.F. Tiersten, Electroelastic interactions and the piezoelectric equations. J. Acoust. Soc. Am. **70**(6), 1567–1576 (1981)

Chapter 5
Finite Element Formulations

Abstract In this chapter, resultant strain and stress are introduced, such that the volume integration can be treat as surface integration. In order to describe the unrestricted finite rotations in thin-walled smart structures, five mechanical nodal DOFs are defined to represent the six kinematic parameters in strain-displacement relations by using Euler angles. Furthermore, an eight-node elements with five mechanical nodal DOFs and additionally integrated with one electrical DOF using full integration or uniformly reduced integration scheme are developed for both composite and smart structures. Implementing both linear constitutive equations and electroelastic nonlinear constitutive equations, one obtains nonlinear FE models by Hamilton's principle and the principle of virtual work, in which various geometrically nonlinear phenomena discussed in Chap. 3 are considered. In the last part of this chapter, several numerical algorithms are developed for solving the nonlinear equilibrium equations and the equations of motion.

5.1 Resultant Vectors

In order to reduce the volume integral to a surface integral in the variational formulation, we define the resultant internal forces and moments per unit length, which can be defined as [1]

$$\overset{n}{L}{}^{\alpha\beta} = \int_{\mathrm{h}} \mu\,(\Theta^3)^n \sigma^{\alpha\beta}\,\mathrm{d}\Theta^3 \qquad (n = 0, 1, 2)\,, \tag{5.1}$$

$$\overset{n}{L}{}^{\alpha 3} = \int_{\mathrm{h}} \mu\,(\Theta^3)^n \sigma^{\alpha 3}\,\mathrm{d}\Theta^3 \qquad (n = 0, 1)\,, \tag{5.2}$$

$$\overset{n}{L}{}^{33} = \int_{\mathrm{h}} \mu\,(\Theta^3)^n \sigma^{33}\,\mathrm{d}\Theta^3 \qquad (n = 0)\,. \tag{5.3}$$

The physical meanings of the resultant internal forces and moments are the in-plane longitudinal forces ($\overset{0}{L}{}^{11}$, $\overset{0}{L}{}^{22}$), the in-plane shear forces ($\overset{0}{L}{}^{12}$, $\overset{0}{L}{}^{21}$), the bending

S.-Q. Zhang, *Nonlinear Analysis of Thin-Walled Smart Structures*, Springer Tracts
in Mechanical Engineering, https://doi.org/10.1007/978-981-15-9857-9_5

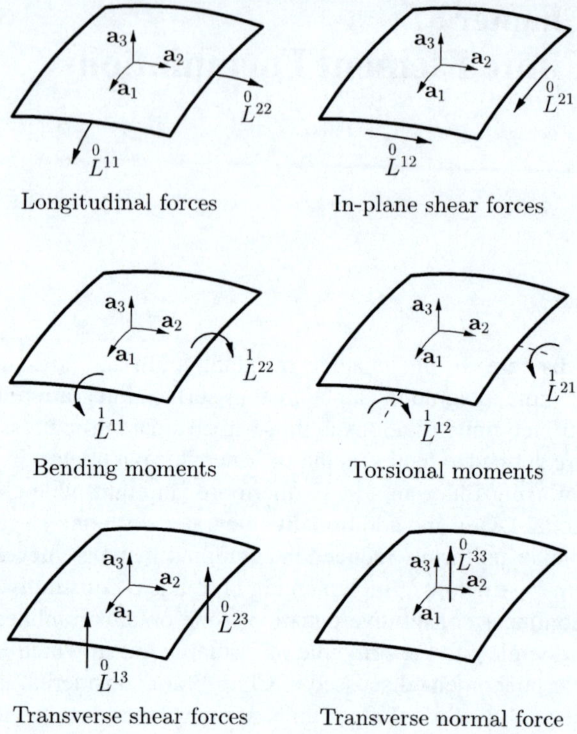

Fig. 5.1 Physical meaning of the resultant internal forces and moments

moments ($\overset{1}{L}^{11}$, $\overset{1}{L}^{22}$), the torsional moments ($\overset{1}{L}^{12}$, $\overset{1}{L}^{21}$), the transverse shear forces ($\overset{0}{L}^{13}$, $\overset{0}{L}^{23}$), and the transverse normal force ($\overset{0}{L}^{33}$), as shown in Fig. 5.1.

The resultant stress vector $\boldsymbol{L}$ and the corresponding resultant strain vector $\boldsymbol{S}$ are defined as

$$\boldsymbol{L} = \left\{ \overset{0}{L}^{11}, \overset{0}{L}^{22}, \overset{0}{L}^{12}, \overset{1}{L}^{11}, \overset{1}{L}^{22}, \overset{1}{L}^{12}, \overset{2}{L}^{11}, \overset{2}{L}^{22}, \overset{2}{L}^{12}, \overset{0}{L}^{23}, \overset{0}{L}^{13}, \overset{1}{L}^{23}, \overset{1}{L}^{13} \right\}^{\mathrm{T}} , \tag{5.4}$$

$$\boldsymbol{S} = \left\{ \overset{0}{\varepsilon}_{11}, \overset{0}{\varepsilon}_{22}, 2\overset{0}{\varepsilon}_{12}, \overset{1}{\varepsilon}_{11}, \overset{1}{\varepsilon}_{22}, 2\overset{1}{\varepsilon}_{12}, \overset{2}{\varepsilon}_{11}, \overset{2}{\varepsilon}_{22}, 2\overset{2}{\varepsilon}_{12}, 2\overset{0}{\varepsilon}_{23}, 2\overset{0}{\varepsilon}_{13}, 2\overset{1}{\varepsilon}_{23}, 2\overset{1}{\varepsilon}_{13} \right\}^{\mathrm{T}} . \tag{5.5}$$

Therefore, the strain components given in (3.68)–(3.70) can be expressed in terms of the resultant strain vector $\boldsymbol{S}$ as

$$\varepsilon = \boldsymbol{H}_{\mathrm{s}} \boldsymbol{S} , \tag{5.6}$$

with

$$
\boldsymbol{H}_{\mathrm{s}} = \begin{bmatrix}
1\,0\,0\,\Theta^3 & 0 & 0 & (\Theta^3)^2 & 0 & 0 & 0\,0\,0 & 0 \\
0\,1\,0\,0 & \Theta^3 & 0 & 0 & (\Theta^3)^2 & 0 & 0\,0\,0 & 0 \\
0\,0\,1\,0 & 0 & \Theta^3 & 0 & 0 & (\Theta^3)^2 & 0\,0\,0 & 0 \\
0\,0\,0\,0 & 0 & 0 & 0 & 0 & 0 & 1\,0\,\Theta^3 & 0 \\
0\,0\,0\,0 & 0 & 0 & 0 & 0 & 0 & 0\,1\,0 & \Theta^3
\end{bmatrix} .
$$

Here the matrix $\boldsymbol{H}_{\mathrm{s}}$ includes the parameter of Θ^3. By this solution, the integral can be first taken through the thickness, leaving only the integral of the in-plane parameters. Using Eqs. (5.4) and (5.5), the volume integral of the internal virtual work can be transformed to a surface integral as

$$
\int_{\mathrm{V}} \delta\varepsilon^{\mathrm{T}}\boldsymbol{\sigma}\,\mathrm{dV} = \int_{\Omega} \delta\boldsymbol{S}^{\mathrm{T}}\boldsymbol{L}\,\mathrm{d}\Omega\,. \tag{5.7}
$$

For later use, we define the vectors $\boldsymbol{v}$ and $\hat{\boldsymbol{v}}_{\mathrm{u}}$ that only contain the generalized displacements and the DOFs, respectively, as

$$
\boldsymbol{v} = \left\{\overset{0}{v}_1\ \overset{0}{v}_2\ \overset{0}{v}_3\ \overset{1}{v}_1\ \overset{1}{v}_2\ \overset{1}{v}_3\right\}^{\mathrm{T}}, \tag{5.8}
$$

$$
\hat{\boldsymbol{v}}_{\mathrm{u}} = \left\{u\ v\ w\ \varphi_1\ \varphi_2\right\}^{\mathrm{T}}. \tag{5.9}
$$

5.2 Rotation Description

The linear shell theory (LIN5) and simplified nonlinear shell theories (RVK5, MRT5, LRT5) have five parameters, while the large rotation nonlinear shell theory (LRT56) has six parameters. All these parameters are the components of displacement vector, usually called generalized displacements. In finite element analysis, they must be expressed by predefined nodal DOFs that have specific physical meanings. In plates and shells, the rotation about Θ_3-axis is compressed, resulting in five nodal DOFs. These five nodal DOFs are composed of three translational DOFs, u, v, w, and two rotational DOFs, φ_1, φ_2, as shown in Fig. 5.2. Here, u, v, w are the translational displacement along the Θ^1-, Θ^2- and Θ^3-axis, respectively, and φ_1, φ_2 are the rotations about the Θ^2- and Θ^1-line, respectively.

The first three parameters, $\overset{0}{v}_1$, $\overset{0}{v}_2$, $\overset{0}{v}_3$, for all shell theories in Chap. 3 can be expressed linearly by the three translational DOFs as

$$
\overset{0}{v}_1 = \frac{\hat{\overset{0}{v}}_1}{\|\boldsymbol{a}^1\|} = \frac{u}{\|\boldsymbol{a}^1\|}, \quad \overset{0}{v}_2 = \frac{\hat{\overset{0}{v}}_2}{\|\boldsymbol{a}^2\|} = \frac{v}{\|\boldsymbol{a}^2\|}, \quad \overset{0}{v}_3 = \hat{\overset{0}{v}}_3 = w\,. \tag{5.10}
$$

The coefficients are generated due to non-unit base vectors used in the development of strain-displacement relations.

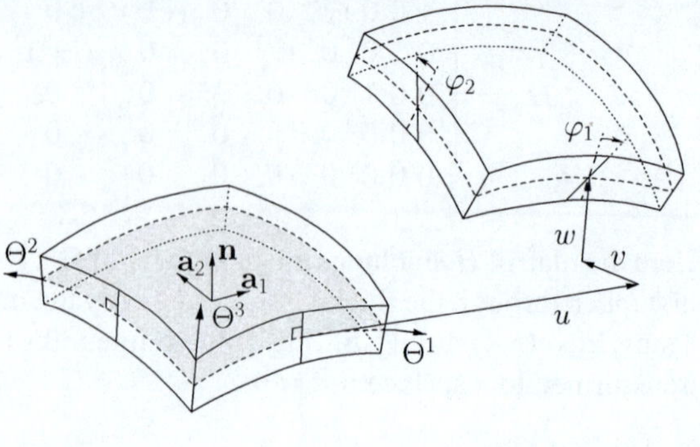

Fig. 5.2 Degrees of freedom at any point on the mid-surface

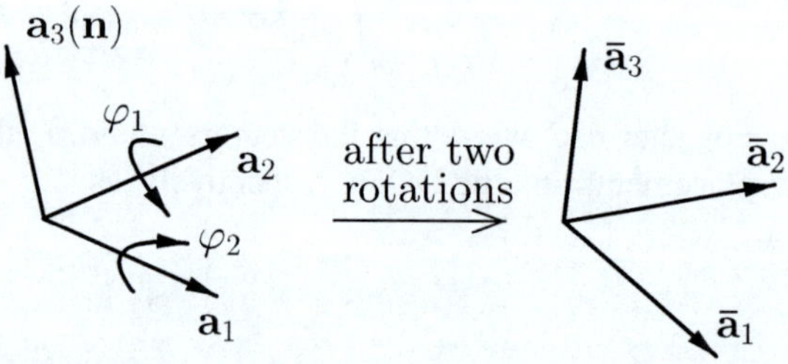

Fig. 5.3 Rotation of the base vector triad by Euler angles φ_1 and φ_2

In LRT56 theory, there are six parameters. The last three parameters of LRT56 theory, $\overset{1}{v}_1$, $\overset{1}{v}_2$, $\overset{1}{v}_3$, should be expressed nonlinearly by two rotational DOFs by using the Euler angle representation, see [2–5]. In order to obtain the mapping matrix between the generalized rotational parameters of LRT56 and two rotational DOFs, the rotation transformation matrix of coordinate system should be defined. Rotating the in-plane coordinate axes sequently by φ_1 about the Θ^2-axis and φ_2 about the Θ^1-axis, as shown in Fig. 5.3, yields the shell director being transformed from $\boldsymbol{n}$ in the undeformed configuration to $\bar{\boldsymbol{a}}_3$ in the deformed configuration.

The transformation matrices of the two independent rotations can be obtained respectively as

$$\boldsymbol{R}_{\mathrm{X}} = \begin{bmatrix} 1 & 0 & 0 \\ 0 & \cos(\varphi_2) & \sin(\varphi_2) \\ 0 & -\sin(\varphi_2) & \cos(\varphi_2) \end{bmatrix}, \quad \boldsymbol{R}_{\mathrm{Y}} = \begin{bmatrix} \cos(\varphi_1) & 0 & \sin(\varphi_1) \\ 0 & 1 & 0 \\ -\sin(\varphi_1) & 0 & \cos(\varphi_1) \end{bmatrix}. \tag{5.11}$$

Here, the matrix $\boldsymbol{R}_{\mathrm{X}}$ is produced by rotating φ_2 about the Θ^1-axis, and the matrix $\boldsymbol{R}_{\mathrm{Y}}$ is by rotating φ_1 about the Θ^2-axis. After the two rotations, the total transformation matrix between the coordinates of the undeformed configuration and the deformed configuration is derived as

$$\begin{Bmatrix} \Theta^1 \\ \Theta^2 \\ \Theta^3 \end{Bmatrix} = \boldsymbol{Rot} \begin{Bmatrix} \bar{\Theta}^1 \\ \bar{\Theta}^2 \\ \bar{\Theta}^3 \end{Bmatrix}, \tag{5.12}$$

with

$$\boldsymbol{Rot} = \boldsymbol{R}_{\mathrm{X}} \cdot \boldsymbol{R}_{\mathrm{Y}} = \begin{bmatrix} \cos(\varphi_1) & -\sin(\varphi_1)\sin(\varphi_2) & \sin(\varphi_1)\cos(\varphi_2) \\ 0 & \cos(\varphi_2) & \sin(\varphi_2) \\ -\sin(\varphi_1) & -\cos(\varphi_1)\sin(\varphi_2) & \cos(\varphi_1)\cos(\varphi_2) \end{bmatrix}, \quad (5.13)$$

$$\boldsymbol{Rot}^{-1} = \begin{bmatrix} \cos(\varphi_1) & 0 & -\sin(\varphi_1) \\ -\sin(\varphi_1)\sin(\varphi_2) & \cos(\varphi_2) & -\cos(\varphi_1)\sin(\varphi_2) \\ \sin(\varphi_1)\cos(\varphi_2) & \sin(\varphi_2) & \cos(\varphi_1)\cos(\varphi_2) \end{bmatrix}. \quad (5.14)$$

Therefore, after two rotations, the covariant base vector in thickness direction of the deformed configuration can be expressed as [4]

$$\bar{\boldsymbol{a}}_3 = \sin(\varphi_1)\cos(\varphi_2)\,\frac{\boldsymbol{a}^1}{\|\boldsymbol{a}^1\|} + \sin(\varphi_2)\,\frac{\boldsymbol{a}^2}{\|\boldsymbol{a}^2\|} + \cos(\varphi_1)\cos(\varphi_2)\,\boldsymbol{a}^3. \quad (5.15)$$

From the definition of the rotational displacement vector, $\overset{1}{\boldsymbol{u}} = \bar{\boldsymbol{a}}_3 - \boldsymbol{n}$, one obtains

$$\begin{aligned} \overset{1}{\boldsymbol{u}} &= \bar{\boldsymbol{a}}_3 - \boldsymbol{n} \\ &= \sin(\varphi_1)\cos(\varphi_2)\,\frac{\boldsymbol{a}^1}{\|\boldsymbol{a}^1\|} + \sin(\varphi_2)\,\frac{\boldsymbol{a}^2}{\|\boldsymbol{a}^2\|} + (\cos(\varphi_1)\cos(\varphi_2) - 1)\,\boldsymbol{a}^3. \end{aligned} \quad (5.16)$$

Thus, the generalized rotational displacements are expressed by two rotational DOFs as

$$\begin{aligned} \overset{1}{v}_1 &= \frac{1}{\|\boldsymbol{a}^1\|}\sin(\varphi_1)\cos(\varphi_2), \\ \overset{1}{v}_2 &= \frac{1}{\|\boldsymbol{a}^2\|}\sin(\varphi_2), \\ \overset{1}{v}_3 &= \cos(\varphi_1)\cos(\varphi_2) - 1. \end{aligned} \quad (5.17)$$

For the linear theory (LIN5), only small rotations are assumed, while, for the simplified nonlinear shell theories (RVK5, MRT5, LRT5), moderate rotations are permitted in structures. The small rotations are defined by $\varphi_\alpha \ll 1$ and the moderate rotations assume that $\varphi_\alpha^2 \ll 1$. Both of these two cases yield $\sin(\varphi_\alpha) = \varphi_\alpha$ and $\cos(\varphi_\alpha) = 1$. Therefore, the generalized rotational displacements for the linear and simplified nonlinear shell theories are approximated as

$$\overset{1}{v}_1 = \frac{1}{\|\boldsymbol{a}^1\|}\varphi_1, \quad \overset{1}{v}_2 = \frac{1}{\|\boldsymbol{a}^2\|}\varphi_2, \quad \overset{1}{v}_3 = 0. \quad (5.18)$$

The generalized rotational displacements of LRT56 theory are expressed nonlinearly by two rotational DOFs. For FE implementation, the nonlinear expressions given in (5.17) must be linearized by means of the Taylor series expansion, with the higher-order terms neglected, as [4]

$$\Delta \overset{1}{v}_i = \left.\frac{\partial \overset{1}{v}_i}{\partial \varphi_1}\right|_t \Delta\varphi_1 + \left.\frac{\partial \overset{1}{v}_i}{\partial \varphi_2}\right|_t \Delta\varphi_2\,, \tag{5.19}$$

where Δ represents the incremental operator. Therefore, the increment of the generalized displacements can be organized in matrix form as

$$\begin{Bmatrix} \Delta \overset{0}{v}_1 \\ \Delta \overset{0}{v}_2 \\ \Delta \overset{0}{v}_3 \\ \Delta \overset{1}{v}_1 \\ \Delta \overset{1}{v}_2 \\ \Delta \overset{1}{v}_3 \end{Bmatrix} = \underbrace{\begin{bmatrix} \frac{1}{\|\boldsymbol{a}^1\|} & 0 & 0 & 0 & 0 \\ 0 & \frac{1}{\|\boldsymbol{a}^2\|} & 0 & 0 & 0 \\ 0 & 0 & 1 & 0 & 0 \\ 0 & 0 & 0 & \frac{\cos(\varphi_1)\cos(\varphi_2)}{\|\boldsymbol{a}^1\|} & \frac{-\sin(\varphi_1)\sin(\varphi_2)}{\|\boldsymbol{a}^1\|} \\ 0 & 0 & 0 & 0 & \frac{\cos(\varphi_2)}{\|\boldsymbol{a}^2\|} \\ 0 & 0 & 0 & -\sin(\varphi_1)\cos(\varphi_2) & -\cos(\varphi_1)\sin(\varphi_2) \end{bmatrix}}_{\boldsymbol{T}_v} \begin{Bmatrix} \Delta u \\ \Delta v \\ \Delta w \\ \Delta\varphi_1 \\ \Delta\varphi_2 \end{Bmatrix}. \tag{5.20}$$

Here, $\boldsymbol{T}_v$ is a transformation matrix of linearization. Thus, the incremental displacement vector $\Delta\boldsymbol{v}$ can be obtained as

$$\Delta\boldsymbol{v} = \boldsymbol{T}_v \Delta\boldsymbol{v}_u\,. \tag{5.21}$$

5.3 Shell Element Design

The whole structures are usually large and with complex geometries. The main concept of finite element analysis is discretizing the structure into small elements. For thin-walled or laminated structures, shell elements are preferred. One of the most popular shell elements is quadrilateral element, which can be classified into Lagrange or Serendipity element, as shown in Fig. 5.4. More detailed description of these two shell elements can be found in most FE books, e.g. Bathe [6], Zienkiewicz et al. [7], Kreja [8]. The elements with quadratic shape functions of both Lagrange and Serendipity elements perform similarly. However, Serendipity elements have less nodes that will save computation time.

Introducing the Jacobian matrix $\boldsymbol{J}$

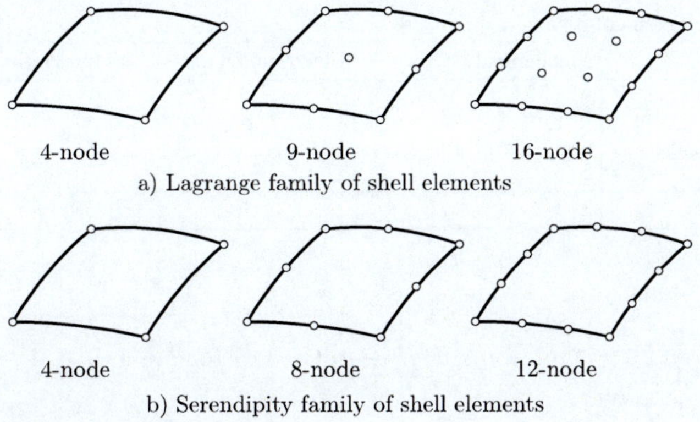

Fig. 5.4 Lagrange and Serendipity families of shell elements

Fig. 5.5 Element mapping between natural coordinates and curvilinear coordinates

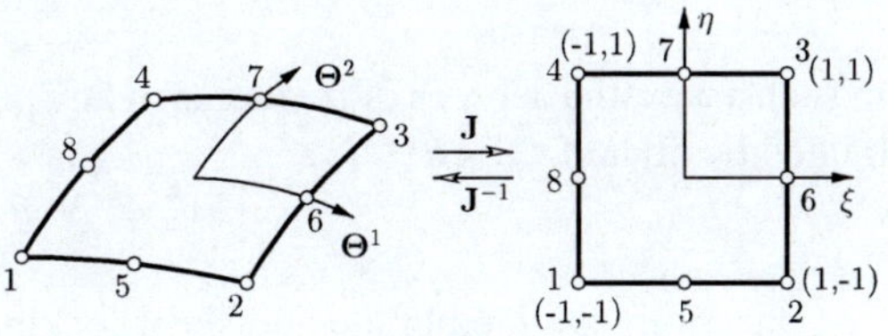

$$\begin{Bmatrix} \dfrac{\partial}{\partial \xi} \\ \dfrac{\partial}{\partial \eta} \end{Bmatrix} = \begin{bmatrix} \dfrac{\partial \Theta^1}{\partial \xi} & \dfrac{\partial \Theta^2}{\partial \xi} \\ \dfrac{\partial \Theta^1}{\partial \eta} & \dfrac{\partial \Theta^2}{\partial \eta} \end{bmatrix} \begin{Bmatrix} \dfrac{\partial}{\partial \Theta^1} \\ \dfrac{\partial}{\partial \Theta^2} \end{Bmatrix} = \boldsymbol{J} \begin{Bmatrix} \dfrac{\partial}{\partial \Theta^1} \\ \dfrac{\partial}{\partial \Theta^2} \end{Bmatrix}, \tag{5.22}$$

the derivatives with respect to the natural coordinates (ξ, η) can connect to the curvilinear coordinates, as shown in Fig. 5.5. Re-writing the formulation in an inverse way, one obtains the transformation relations from the curvilinear coordinates to the natural coordinates as

$$\begin{Bmatrix} \dfrac{\partial}{\partial \Theta^1} \\ \dfrac{\partial}{\partial \Theta^2} \end{Bmatrix} = \boldsymbol{J}^{-1} \begin{Bmatrix} \dfrac{\partial}{\partial \xi} \\ \dfrac{\partial}{\partial \eta} \end{Bmatrix} = \frac{1}{|\boldsymbol{J}|} \begin{bmatrix} \dfrac{\partial \Theta^2}{\partial \eta} & -\dfrac{\partial \Theta^2}{\partial \xi} \\ -\dfrac{\partial \Theta^1}{\partial \eta} & \dfrac{\partial \Theta^1}{\partial \xi} \end{bmatrix} \begin{Bmatrix} \dfrac{\partial}{\partial \xi} \\ \dfrac{\partial}{\partial \eta} \end{Bmatrix}, \tag{5.23}$$

which is frequently used in the FE modeling, since the equations of strain-displacement relations are always given in curvilinear coordinates.

In the present study, the eight-node Serendipity shell element is considered. The interpolation functions, usually called shape functions, can be expressed at each node in the natural coordinate system as

Table 5.1 Shell element types

Element	Mechanical DOFs	Electrical DOFs	Integration scheme
SH85FI	5	0	FI
SH85URI	5	0	URI
SH851FI	5	1	FI
SH851URI	5	1	URI

$$
\begin{aligned}
N_I &= \frac{1}{4}(1+\xi_I\xi)(1+\eta_I\eta)(\xi_I\xi+\eta_I\eta-1) && \text{for } I \in 1,2,3,4\,,\\
N_I &= \frac{1}{2}(1-\xi^2)(1+\eta_I\eta) && \text{for } I \in 5,7\,,\\
N_I &= \frac{1}{2}(1-\eta^2)(1+\xi_I\xi) && \text{for } I \in 6,8\,.
\end{aligned}
\tag{5.24}
$$

In such a way, the degrees of freedoms of any point at the mid-surface can be approximated by nodal DOFs $\boldsymbol{q}$

$$
\boldsymbol{v}_{\mathrm{u}} = \boldsymbol{N}_{\mathrm{u}}\boldsymbol{q}\,. \tag{5.25}
$$

Concerning the membrane and shear locking problems, several numerical methods, e.g. ANS, EAS, SRI or URI, have been mentioned in Chap. 2. In this report, only the URI scheme is employed to avoid shear locking. For comparison, the FI scheme is used in some examples.

Two abbreviations of elements, SH85FI and SH85URI, are defined for composite structures. They denote eight-node isoparametric shell elements with five mechanical nodal DOFs using respectively FI and URI integration schemes. In addition, two piezoelectric coupled elements denoted as SH851FI and SH851URI are defined. They represent eight-node isoparametric shell elements with five mechanical nodal DOFs and one electrical DOF per piezoelectric material layer respectively using FI and URI integration schemes. All the shell elements used in the later simulations are listed in Table 5.1.

5.4 Variational Formulations

In order to derive the dynamic equations of composite or laminated smart structures, Hamilton's principle is employed, which is defined by

$$
\int_{t_1}^{t_2} \left(\delta T - \delta W_{\mathrm{int}} + \delta W_{\mathrm{ext}}\right)\,\mathrm{d}t = 0\,, \tag{5.26}
$$

where δ represents the variational operator, T, W_{int} and W_{ext} are the kinetic energy, the internal work and the external work, respectively. For static equilibrium equation of smart structures, the principle of virtual work is employed, which is given by

$$\delta W_{\text{int}} = \delta W_{\text{ext}} \, . \tag{5.27}$$

The variation of the kinetic energy, δT, can be calculated by [9]

$$\delta T = \int_{\text{V}} \rho \, \delta \dot{\boldsymbol{u}}^{\text{T}} \dot{\boldsymbol{u}} \, \text{dV} = - \int_{\text{V}} \rho \, \delta \boldsymbol{u}^{\text{T}} \ddot{\boldsymbol{u}} \, \text{dV} \, , \tag{5.28}$$

where ρ is the material density, $\dot{\square}$ and $\ddot{\square}$ denote respectively the first- and second-order time derivative. Furthermore, $\boldsymbol{u}$ denotes the vector of the displacements in the shell space, which is given by

$$\boldsymbol{u} = \begin{bmatrix} 1 & 0 & 0 & \Theta^3 & 0 & 0 \\ 0 & 1 & 0 & 0 & \Theta^3 & 0 \\ 0 & 0 & 1 & 0 & 0 & \Theta^3 \end{bmatrix} \begin{Bmatrix} \overset{0}{v}_1 \\ \overset{0}{v}_2 \\ \overset{0}{v}_3 \\ \overset{1}{v}_1 \\ \overset{1}{v}_2 \\ \overset{1}{v}_3 \end{Bmatrix} = \boldsymbol{Z}_{\text{u}} \boldsymbol{v} \, , \tag{5.29}$$

where $\boldsymbol{v}$ is the generalized displacement vector.

According to (5.28) and (5.29), δT can be written as

$$\delta T = - \int_{\text{V}} \rho \, \delta \boldsymbol{v}^{\text{T}} \boldsymbol{Z}_{\text{u}}^{\text{T}} \boldsymbol{Z}_{\text{u}} \ddot{\boldsymbol{v}} \, \text{dV} = - \int_{\Omega} \delta \boldsymbol{v}^{\text{T}} \boldsymbol{H}_{\text{u}} \ddot{\boldsymbol{v}} \, \text{d}\Omega \, , \tag{5.30}$$

in which

$$\boldsymbol{H}_{\text{u}} = \int_{h} \rho \, \boldsymbol{Z}_{\text{u}}^{\text{T}} \boldsymbol{Z}_{\text{u}} \, \mu \, \text{d}\Theta^3 \, . \tag{5.31}$$

The variation of the potential energy or internal virtual work, δW_{int}, is given by

$$\delta W_{\text{int}} = \int_{\text{V}} \left(\delta \boldsymbol{\varepsilon}^{\text{T}} \boldsymbol{\sigma} - \delta \boldsymbol{E}^{\text{T}} \boldsymbol{D} \right) \text{dV} \, . \tag{5.32}$$

Inserting constitutive equations into (5.32) yields

$$\begin{aligned} \delta W_{\text{int}} &= \int_{\text{V}} \left(\delta \boldsymbol{\varepsilon}^{\text{T}} \boldsymbol{c} \boldsymbol{\varepsilon} - \delta \boldsymbol{\varepsilon}^{\text{T}} \boldsymbol{e}^{\text{T}} \boldsymbol{E} - \delta \boldsymbol{E}^{\text{T}} \boldsymbol{e} \boldsymbol{\varepsilon} - \delta \boldsymbol{E}^{\text{T}} \boldsymbol{\epsilon} \boldsymbol{E} \right) \text{dV} \\ &= \delta W_{\text{int}}^{(1)} + \delta W_{\text{int}}^{(2)} + \delta W_{\text{int}}^{(3)} + \delta W_{\text{int}}^{(4)} \, , \end{aligned} \tag{5.33}$$

where $\delta W_{\text{int}}^{(1)}$ and $\delta W_{\text{int}}^{(2)}$ are the pure and piezoelectric coupled mechanical internal virtual work, while $\delta W_{\text{int}}^{(3)}$ and $\delta W_{\text{int}}^{(4)}$ represent the coupled and pure electrical internal virtual work, respectively.

By using the resultant strain and stress vectors, $\delta W_{\text{int}}^{(1)}$, $\delta W_{\text{int}}^{(2)}$, $\delta W_{\text{int}}^{(3)}$, $\delta W_{\text{int}}^{(4)}$, given in (5.33), can be organized as

$$\delta W_{\text{int}}^{(1)} = \int_{\text{V}} \delta \boldsymbol{\varepsilon}^{\text{T}} \boldsymbol{c} \boldsymbol{\varepsilon} \, \text{dV} = \int_{\Omega} \delta \boldsymbol{S}^{\text{T}} \boldsymbol{H}_{\text{c}} \boldsymbol{S} \, \text{d}\Omega \,, \tag{5.34}$$

$$\delta W_{\text{int}}^{(2)} = - \int_{\text{V}} \delta \boldsymbol{\varepsilon}^{\text{T}} \boldsymbol{e}^{\text{T}} \boldsymbol{E} \, \text{dV} = \int_{\Omega} \delta \boldsymbol{S}^{\text{T}} \boldsymbol{H}_{\text{e}}^{\text{T}} \boldsymbol{E} \, \text{d}\Omega \,, \tag{5.35}$$

$$\delta W_{\text{int}}^{(3)} = - \int_{\text{V}} \delta \boldsymbol{E}^{\text{T}} \boldsymbol{e} \boldsymbol{\varepsilon} \, \text{dV} = \int_{\Omega} \delta \boldsymbol{E}^{\text{T}} \boldsymbol{H}_{\text{e}} \boldsymbol{S} \, \text{d}\Omega \,, \tag{5.36}$$

$$\delta W_{\text{int}}^{(4)} = - \int_{\text{V}} \delta \boldsymbol{E}^{\text{T}} \boldsymbol{\epsilon} \boldsymbol{E} \, \text{dV} = \int_{\Omega} \delta \boldsymbol{E}^{\text{T}} \boldsymbol{H}_{\text{g}} \boldsymbol{E} \, \text{d}\Omega \,, \tag{5.37}$$

with

$$\boldsymbol{H}_{\text{c}} = \int_{\text{h}} \boldsymbol{H}_{\text{s}}^{\text{T}} \boldsymbol{c} \boldsymbol{H}_{\text{s}} \, \mu \, \text{d}\Theta^3 \,, \tag{5.38}$$

$$\boldsymbol{H}_{\text{e}} = - \int_{\text{h}} \boldsymbol{e} \boldsymbol{H}_{\text{s}} \, \mu \, \text{d}\Theta^3 \,, \tag{5.39}$$

$$\boldsymbol{H}_{\text{g}} = - \int_{\text{h}} \boldsymbol{\epsilon} \, \mu \, \text{d}\Theta^3 \,. \tag{5.40}$$

Furthermore, the external virtual work, δW_{ext}, can be derived as [9, 10]

$$\delta W_{\text{ext}} = \int_{\text{V}} \delta \boldsymbol{u}^{\text{T}} \boldsymbol{f}_{\text{b}} \, \text{dV} + \int_{\Omega} \delta \boldsymbol{u}^{\text{T}} \boldsymbol{f}_{\text{s}} \, \text{d}\Omega + \delta \boldsymbol{u}^{\text{T}} \boldsymbol{f}_{\text{c}} - \int_{\Omega} \delta \boldsymbol{\phi}^{\text{T}} \boldsymbol{\varrho} \, \text{d}\Omega - \delta \boldsymbol{\phi}^{\text{T}} \boldsymbol{Q}_{\text{c}} \,, \tag{5.41}$$

where $\boldsymbol{f}_{\text{b}}$, $\boldsymbol{f}_{\text{s}}$ and $\boldsymbol{f}_{\text{c}}$ denote the body force, the surface distributed force and the concentrated force vectors, associated with base vectors of curvilinear coordinate axes. Additionally, $\boldsymbol{\varrho}$ is the surface charge vector and $\boldsymbol{Q}_{\text{c}}$ the applied concentrated electric charge vector.

5.5 Total Lagrangian Formulation

For linearization of nonlinear FE equations, total Lagrangian (TL) incremental formulation [4, 10–12] are adopted. Three configurations are defined and considered for structures, listed in Table 5.2. The configurations are characterized by the left superscripts 0, 1 or 2, the reference configurations are denoted by the left subscripts 0. Using the TL method, the stress vector, the strain vector, the displacement vector, etc. in the virtual configuration can be expressed by those in the current configuration

Table 5.2 Notations for different configurations

Notation	Meaning
${}^{0}\mathrm{C}$	Initial configuration, referring to the undeformed configuration
${}^{1}\mathrm{C}$	Current configuration, referring to the deformed configuration
${}^{2}\mathrm{C}$	Virtual configuration, which is called searched configuration
${}^{m}\mathrm{C}$	Configuration m, $m = 0, 1, 2,$

and the incremental values as

$$ {}^{2}_{0}\boldsymbol{X} = {}^{1}_{0}\boldsymbol{X} + \Delta\boldsymbol{X}, \quad (\boldsymbol{X} = \boldsymbol{L}, \boldsymbol{S}, \boldsymbol{D}, \boldsymbol{E}, \boldsymbol{v}, \boldsymbol{\phi}) \ . \tag{5.42} $$

The strain components of geometrically nonlinear theories are composed of higher-order terms of generalized displacements. For linearization procedure, the increment of resultant strain vector can be derived as

$$ \Delta\boldsymbol{S} = \begin{bmatrix} \dfrac{\partial S_1}{\partial \overset{0}{v}_1} & \dfrac{\partial S_1}{\partial \overset{0}{v}_2} & \cdots & \dfrac{\partial S_1}{\partial \overset{1}{v}_3} \\ \dfrac{\partial S_2}{\partial \overset{0}{v}_1} & \dfrac{\partial S_2}{\partial \overset{0}{v}_2} & \cdots & \dfrac{\partial S_2}{\partial \overset{1}{v}_3} \\ \vdots & \vdots & \ddots & \vdots \\ \dfrac{\partial S_{13}}{\partial \overset{0}{v}_1} & \dfrac{\partial S_{13}}{\partial \overset{0}{v}_2} & \cdots & \dfrac{\partial S_{13}}{\partial \overset{1}{v}_3} \end{bmatrix} \begin{Bmatrix} \Delta\overset{0}{v}_1 \\ \Delta\overset{0}{v}_2 \\ \Delta\overset{0}{v}_3 \\ \Delta\overset{1}{v}_1 \\ \Delta\overset{1}{v}_2 \\ \Delta\overset{1}{v}_3 \end{Bmatrix} = \boldsymbol{A}_{\mathrm{s}} \cdot \Delta\boldsymbol{v}. \tag{5.43} $$

Using the relations given in Eqs. (5.21) and (5.25), one obtains

$$ \Delta\boldsymbol{S} = \boldsymbol{A}_{\mathrm{s}}\boldsymbol{T}_{\mathrm{v}}\boldsymbol{N}_{\mathrm{u}}\Delta\boldsymbol{q} = \boldsymbol{B}_{\mathrm{u}}\Delta\boldsymbol{q}. \tag{5.44} $$

where $\boldsymbol{B}_{\mathrm{u}}$ denotes the linearized strain field matrix.

From Eq. (5.30), the variation of the kinetic energy in the virtual configuration, ${}^{2}_{0}\delta T$, can be obtained as

$$ \begin{aligned} {}^{2}_{0}\delta T &= -\int_{\Omega} {}^{2}_{0}\delta\boldsymbol{v}^{\mathrm{T}}\boldsymbol{H}_{\mathrm{u}}{}^{2}_{0}\ddot{\boldsymbol{v}}\,\mathrm{d}\Omega \\ &= -\,\delta\Delta\boldsymbol{q}^{\mathrm{T}}\Big(\int_{\Omega} \boldsymbol{N}_{\mathrm{u}}^{\mathrm{T}}\boldsymbol{T}_{\boldsymbol{v}}^{\mathrm{T}}\boldsymbol{H}_{\mathrm{u}}{}^{1}_{0}\ddot{\boldsymbol{v}}\,\mathrm{d}\Omega + \int_{\Omega} \boldsymbol{N}_{\mathrm{u}}^{\mathrm{T}}\boldsymbol{T}_{\boldsymbol{v}}^{\mathrm{T}}\boldsymbol{H}_{\mathrm{u}}\boldsymbol{T}_{\boldsymbol{v}}\boldsymbol{N}_{\mathrm{u}}\,\mathrm{d}\Omega\,\Delta\ddot{\boldsymbol{q}}\Big) \\ &= -\,\delta\Delta\boldsymbol{q}^{\mathrm{T}}\big({}^{1}\boldsymbol{F}_{\mathrm{ut}} + {}^{1}\boldsymbol{M}_{\mathrm{uu}}\Delta\ddot{\boldsymbol{q}}\big), \end{aligned} \tag{5.45} $$

where ${}^{1}\boldsymbol{F}_{\mathrm{ut}}$ and ${}^{1}\boldsymbol{M}_{\mathrm{uu}}$ represent the inertial in-balance force and mass matrix, which are respectively calculated by

$$ {}^{1}\boldsymbol{F}_{\mathrm{ut}} = \int_{\Omega} \boldsymbol{N}_{\mathrm{u}}^{\mathrm{T}} \boldsymbol{T}_{\boldsymbol{v}}^{\mathrm{T}} \boldsymbol{H}_{\mathrm{u}0}{}^{1}\ddot{\boldsymbol{v}} \, \mathrm{d}\Omega \,, \tag{5.46} $$

$$ {}^{1}\boldsymbol{M}_{\mathrm{uu}} = \int_{\Omega} \boldsymbol{N}_{\mathrm{u}}^{\mathrm{T}} \boldsymbol{T}_{\boldsymbol{v}}^{\mathrm{T}} \boldsymbol{H}_{\mathrm{u}} \boldsymbol{T}_{\boldsymbol{v}} \boldsymbol{N}_{\mathrm{u}} \, \mathrm{d}\Omega \,. \tag{5.47} $$

From Eq. (5.34), the pure mechanical induced virtual work in the virtual configuration, ${}_{0}^{2}\delta W_{\mathrm{int}}^{(1)}$, can be expressed as

$$ \begin{aligned} {}_{0}^{2}\delta W_{\mathrm{int}}^{(1)} &= \int_{\Omega} {}_{0}^{2}\delta \boldsymbol{S}^{\mathrm{T}} \boldsymbol{H}_{\mathrm{c}0}{}^{2}\boldsymbol{S} \, \mathrm{d}\Omega \\ &= \delta \Delta \boldsymbol{q}^{\mathrm{T}} \left(\int_{\Omega} \boldsymbol{B}_{\mathrm{u}}^{\mathrm{T}} \boldsymbol{H}_{\mathrm{c}0}{}^{1}\boldsymbol{S} \, \mathrm{d}\Omega + \int_{\Omega} \boldsymbol{B}_{\mathrm{u}}^{\mathrm{T}} \boldsymbol{H}_{\mathrm{c}} \boldsymbol{B}_{\mathrm{u}} \mathrm{d}\Omega \, \Delta \boldsymbol{q} \right) \\ &= \delta \Delta \boldsymbol{q}^{\mathrm{T}} \left({}^{1}\boldsymbol{F}_{\mathrm{uu}} + {}^{1}\boldsymbol{K}_{\mathrm{uu}} \Delta \boldsymbol{q} \right) , \end{aligned} \tag{5.48} $$

where ${}^{1}\boldsymbol{F}_{\mathrm{uu}}$ and ${}^{1}\boldsymbol{K}_{\mathrm{uu}}$ denote the mechanically induced in-balance force vector and the linearized stiffness matrix, respectively. The linearized and geometrically nonlinear stiffness matrices will be updated after every iteration. The mechanically induced in-balance force vector ${}^{1}\boldsymbol{F}_{\mathrm{uu}}$ and the linearized stiffness matrix ${}^{1}\boldsymbol{K}_{\mathrm{uu}}$ can be respectively obtained as

$$ {}^{1}\boldsymbol{F}_{\mathrm{uu}} = \int_{\Omega} \boldsymbol{B}_{\mathrm{u}}^{\mathrm{T}} \boldsymbol{H}_{\mathrm{c}0}{}^{1}\boldsymbol{S} \, \mathrm{d}\Omega \,, \tag{5.49} $$

$$ {}^{1}\boldsymbol{K}_{\mathrm{uu}} = \int_{\Omega} \boldsymbol{B}_{\mathrm{u}}^{\mathrm{T}} \boldsymbol{H}_{\mathrm{c}} \boldsymbol{B}_{\mathrm{u}} \, \mathrm{d}\Omega \,. \tag{5.50} $$

From Eq. (5.35), the coupled mechanical internal virtual work in the virtual configuration, ${}_{0}^{2}\delta W_{\mathrm{int}}^{(2)}$, can be expressed as

$$ \begin{aligned} {}_{0}^{2}\delta W_{\mathrm{int}}^{(2)} &= \int_{\Omega} {}_{0}^{2}\delta \boldsymbol{S}^{\mathrm{T}} \boldsymbol{H}_{\mathrm{e}0}^{\mathrm{T}2}\boldsymbol{E} \, \mathrm{d}\Omega \\ &= \delta \Delta \boldsymbol{q}^{\mathrm{T}} \left(\int_{\Omega} \boldsymbol{B}_{\mathrm{u}}^{\mathrm{T}} \boldsymbol{H}_{\mathrm{e}0}^{\mathrm{T}1}\boldsymbol{E} \, \mathrm{d}\Omega + \int_{\Omega} \boldsymbol{B}_{\mathrm{u}}^{\mathrm{T}} \boldsymbol{H}_{\mathrm{e}}^{\mathrm{T}} \boldsymbol{B}_{\phi} \, \mathrm{d}\Omega \, \Delta \phi \right) \\ &= \delta \Delta \boldsymbol{q}^{\mathrm{T}} \left({}^{1}\boldsymbol{F}_{\mathrm{u}\phi} + {}^{1}\boldsymbol{K}_{\mathrm{u}\phi} \Delta \phi \right) , \end{aligned} \tag{5.51} $$

where ${}^{1}\boldsymbol{F}_{\mathrm{u}\phi}$, ${}^{1}\boldsymbol{K}_{\mathrm{u}\phi}$ are the electrically induced in-balance force vector, the coupled stiffness matrix. They can be calculated by

$$^{1}\boldsymbol{F}_{\mathrm{u}\phi}=\int_{\Omega}\boldsymbol{B}_{\mathrm{u}}^{\mathrm{T}}\boldsymbol{H}_{\mathrm{e}}^{\mathrm{T}}{}_{0}^{1}\boldsymbol{E}\,\mathrm{d}\Omega\,, \tag{5.52}$$

$$^{1}\boldsymbol{K}_{\mathrm{u}\phi}=\int_{\Omega}\boldsymbol{B}_{\mathrm{u}}^{\mathrm{T}}\boldsymbol{H}_{\mathrm{e}}^{\mathrm{T}}\boldsymbol{B}\phi\,\mathrm{d}\Omega\,. \tag{5.53}$$

From Eq. (5.36), the coupled electrical internal virtual work in the virtual configuration, ${}_{0}^{2}\delta W_{\mathrm{int}}^{(3)}$, can be calculated as

$$\begin{aligned}{}_{0}^{2}\delta W_{\mathrm{int}}^{(3)}&=\int_{\Omega}{}_{0}^{2}\delta\boldsymbol{E}^{\mathrm{T}}\boldsymbol{H}_{\mathrm{e}}{}_{0}^{2}\boldsymbol{S}\,\mathrm{d}\Omega\\&=\delta\Delta\boldsymbol{\phi}^{\mathrm{T}}\left(\int_{\Omega}\boldsymbol{B}_{\phi}^{\mathrm{T}}\boldsymbol{H}_{\mathrm{e}}{}_{0}^{1}\boldsymbol{S}\,\mathrm{d}\Omega+\int_{{}^{0}\Omega}\boldsymbol{B}_{\phi}^{\mathrm{T}}\boldsymbol{H}_{\mathrm{e}}\boldsymbol{B}_{\mathrm{u}}\,\mathrm{d}\Omega\,\Delta\boldsymbol{q}\right)\\&=\delta\Delta\boldsymbol{\phi}^{\mathrm{T}}\left({}^{1}\boldsymbol{F}_{\phi\mathrm{u}}+{}^{1}\boldsymbol{K}_{\phi\mathrm{u}}\Delta\boldsymbol{q}\right),\end{aligned} \tag{5.54}$$

Here, ${}^{1}\boldsymbol{F}_{\phi\mathrm{u}}$ and ${}^{1}\boldsymbol{K}_{\phi\mathrm{u}}$ denote the mechanically induced in-balance charge vector and the piezoelectric coupled capacity matrix, which are respectively given by

$$^{1}\boldsymbol{F}_{\phi\mathrm{u}}=\int_{\Omega}\boldsymbol{B}_{\phi}^{\mathrm{T}}\boldsymbol{H}_{\mathrm{e}}{}_{0}^{1}\boldsymbol{S}\,\mathrm{d}\Omega\,, \tag{5.55}$$

$$^{1}\boldsymbol{K}_{\phi\mathrm{u}}=\int_{\Omega}\boldsymbol{B}_{\phi}^{\mathrm{T}}\boldsymbol{H}_{\mathrm{e}}\boldsymbol{B}_{\mathrm{u}}\,\mathrm{d}\Omega\,. \tag{5.56}$$

From Eq. (5.37), the pure electric internal virtual work in the virtual configuration, ${}_{0}^{2}\delta W_{\mathrm{int}}^{(4)}$, can be expressed as

$$\begin{aligned}{}_{0}^{2}\delta W_{\mathrm{int}}^{(4)}&=\int_{\Omega}{}_{0}^{2}\delta\boldsymbol{E}^{\mathrm{T}}\boldsymbol{H}_{\mathrm{g}}{}_{0}^{2}\boldsymbol{E}\,\mathrm{d}\Omega\\&=\delta\Delta\boldsymbol{\phi}^{\mathrm{T}}\left(\int_{\Omega}\boldsymbol{B}_{\phi}^{\mathrm{T}}\boldsymbol{H}_{\mathrm{g}}{}_{0}^{1}\boldsymbol{E}\,\mathrm{d}\Omega+\int_{\Omega}\boldsymbol{B}_{\phi}^{\mathrm{T}}\boldsymbol{H}_{\mathrm{g}}\boldsymbol{B}_{\phi}\,\mathrm{d}\Omega\,\Delta\boldsymbol{\phi}\right)\\&=\delta\Delta\boldsymbol{\phi}^{\mathrm{T}}\left({}^{1}\boldsymbol{F}_{\phi\phi}+{}^{1}\boldsymbol{K}_{\phi\phi}\Delta\boldsymbol{\phi}\right),\end{aligned} \tag{5.57}$$

in which the electrically induced in-balance charge vector ${}^{1}\boldsymbol{F}_{\phi\phi}$ and the piezoelectric capacity matrix ${}^{1}\boldsymbol{K}_{\phi\phi}$ are calculated as

$$^{1}\boldsymbol{F}_{\phi\phi}=\int_{\Omega}\boldsymbol{B}_{\phi}^{\mathrm{T}}\boldsymbol{H}_{\mathrm{g}}{}_{0}^{1}\boldsymbol{E}\,\mathrm{d}\Omega\,, \tag{5.58}$$

$$^{1}\boldsymbol{K}_{\phi\phi}=\int_{\Omega}\boldsymbol{B}_{\phi}^{\mathrm{T}}\boldsymbol{H}_{\mathrm{g}}\boldsymbol{B}_{\phi}\,\mathrm{d}\Omega\,. \tag{5.59}$$

The variation of the external work in the virtual configuration, ${}_{0}^{2}\delta W_{\mathrm{ext}}$, including the mechanical force and electric charge loads, are expressed as

$$
\begin{aligned}
{}_0^2\delta W_{\text{ext}} &= \int_{\text{V}} {}_0^2\delta \boldsymbol{u}^{\text{T}} \boldsymbol{f}_{\text{b}} \,\text{dV} + \int_{\Omega} {}_0^2\delta \boldsymbol{u}^{\text{T}} \boldsymbol{f}_{\text{s}} \,\text{d}\Omega + {}_0^2\delta \boldsymbol{u}^{\text{T}} \boldsymbol{f}_{\text{c}} - \int_{\Omega} {}_0^2\delta \boldsymbol{\phi}^{\text{T}} \boldsymbol{\varrho} \,\text{d}\Omega - {}_0^2\delta \boldsymbol{\phi}^{\text{T}} \boldsymbol{Q}_{\text{c}} \\
&= \delta\Delta \boldsymbol{q}^{\text{T}} \Big(\boldsymbol{F}_{\text{ub}} + \boldsymbol{F}_{\text{us}} + \boldsymbol{F}_{\text{uc}} \Big) + \delta\Delta \boldsymbol{\phi}^{\text{T}} \Big(\boldsymbol{G}_{\phi\text{s}} + \boldsymbol{G}_{\phi\text{c}} \Big) ,
\end{aligned}
\tag{5.60}
$$

with

$$
\boldsymbol{F}_{\text{ub}} = \int_{\text{V}} \boldsymbol{N}_{\text{u}}^{\text{T}} \boldsymbol{T}_{\text{v}}^{\text{T}} \boldsymbol{Z}_{\text{u}}^{\text{T}} \boldsymbol{f}_{\text{b}} \,\text{dV} , \tag{5.61}
$$

$$
\boldsymbol{F}_{\text{us}} = \int_{\Omega} \boldsymbol{N}_{\text{u}}^{\text{T}} \boldsymbol{T}_{\text{v}}^{\text{T}} \boldsymbol{Z}_{\text{u}}^{\text{T}} \boldsymbol{f}_{\text{s}} \,\text{d}\Omega , \tag{5.62}
$$

$$
\boldsymbol{F}_{\text{uc}} = \boldsymbol{N}_{\text{u}}^{\text{T}} \boldsymbol{T}_{\text{v}}^{\text{T}} \boldsymbol{Z}_{\text{u}}^{\text{T}} \boldsymbol{f}_{\text{c}} , \tag{5.63}
$$

$$
\boldsymbol{G}_{\phi\text{s}} = -\int_{\Omega} \boldsymbol{\varrho} \,\text{d}\Omega , \tag{5.64}
$$

$$
\boldsymbol{G}_{\phi\text{c}} = -\boldsymbol{Q}_{\text{c}} , \tag{5.65}
$$

where $\boldsymbol{F}_{\text{ub}}, \boldsymbol{F}_{\text{us}}, \boldsymbol{F}_{\text{uc}}$ are the element body force, surface force and concentrated force vectors, respectively, while $\boldsymbol{G}_{\phi\text{s}}$ and $\boldsymbol{G}_{\phi\phi}$ denote the element surface and concentrated electric charge vectors that are applied on piezoelectric material layers.

5.6 Geometrically Nonlinear FE Models

5.6.1 Dynamic FE Model

Substituting Eqs. (5.45), (5.48), (5.51), (5.54), (5.57) and (5.60) into the Hamilton's principle given in (5.26) yields

$$
\begin{aligned}
0 = {} & \delta\Delta \boldsymbol{q}^{\text{T}} \Big({}^1\boldsymbol{F}_{\text{ut}} + {}^1\boldsymbol{M}_{\text{uu}} \Delta \ddot{\boldsymbol{q}} \Big) \\
& + \delta\Delta \boldsymbol{q}^{\text{T}} \Big({}^1\boldsymbol{F}_{\text{uu}} + {}^1\boldsymbol{K}_{\text{uu}} \Delta \boldsymbol{q} + {}^1\boldsymbol{F}_{\text{u}\phi} + {}^1\boldsymbol{K}_{\text{u}\phi} \Delta \boldsymbol{\phi} \Big) \\
& + \delta\Delta \boldsymbol{\phi}^{\text{T}} \Big({}^1\boldsymbol{F}_{\phi\text{u}} + {}^1\boldsymbol{K}_{\phi\text{u}} \Delta \boldsymbol{q} + {}^1\boldsymbol{F}_{\phi\phi} + {}^1\boldsymbol{K}_{\phi\phi} \Delta \boldsymbol{\phi} \Big) \\
& - \delta\Delta \boldsymbol{q}^{\text{T}} \Big(\boldsymbol{F}_{\text{ub}} + \boldsymbol{F}_{\text{us}} + \boldsymbol{F}_{\text{uc}} \Big) \\
& - \delta\Delta \boldsymbol{\phi}^{\text{T}} \Big(\boldsymbol{G}_{\phi\text{s}} + \boldsymbol{G}_{\phi\text{c}} \Big) .
\end{aligned}
\tag{5.66}
$$

In order to satisfy Eq. (5.66) unconditionally, the coefficient terms in front of $\delta\Delta \boldsymbol{q}^{\text{T}}$ and $\delta\Delta \boldsymbol{\phi}^{\text{T}}$ must be set to zero, respectively, which yields a piezoelectric coupled dynamic FE model including an equation of motion and a sensor equation as

$$ {}^{1}\boldsymbol{M}_{\mathrm{uu}0}^{\ 2}\ddot{\boldsymbol{q}} + {}^{1}\boldsymbol{K}_{\mathrm{uu}}\Delta\boldsymbol{q} + {}^{1}\boldsymbol{K}_{\mathrm{u}\phi}\Delta\boldsymbol{\phi}_{a} = \boldsymbol{F}_{\mathrm{ue}} - {}^{1}\boldsymbol{F}_{\mathrm{ui}}\,, \tag{5.67} $$

$$ {}^{1}\boldsymbol{K}_{\phi\mathrm{u}}\Delta\boldsymbol{q} + {}^{1}\boldsymbol{K}_{\phi\phi}\Delta\boldsymbol{\phi}_{s} = \boldsymbol{G}_{\phi\mathrm{e}} - {}^{1}\boldsymbol{G}_{\phi\mathrm{i}}\,, \tag{5.68} $$

where ${}^{1}\boldsymbol{M}_{\mathrm{uu}}$, ${}^{1}\boldsymbol{K}_{\mathrm{uu}}$, ${}^{1}\boldsymbol{K}_{\mathrm{u}\phi}$, ${}^{1}\boldsymbol{K}_{\phi\mathrm{u}}$ and ${}^{1}\boldsymbol{K}_{\phi\phi}$ represent the mass, the total stiffness, the coupled stiffness, the coupled capacity and the piezoelectric capacity matrices, respectively. In the right-hand side of the above equations, $\boldsymbol{F}_{\mathrm{ue}}$, ${}^{1}\boldsymbol{F}_{\mathrm{ui}}$, $\boldsymbol{G}_{\phi\mathrm{e}}$ and ${}^{1}\boldsymbol{G}_{\phi\mathrm{i}}$ denote the external force, the in-balance force, the external charge and the in-balance charge vectors, respectively. Additionally, $\ddot{\boldsymbol{q}}$ is the acceleration of the nodal DOF vector, $\boldsymbol{q}$ the nodal DOF vector, $\boldsymbol{\phi}_{\mathrm{a}}$ the vector of the electric potential applied on piezoelectric material layers, and $\boldsymbol{\phi}_{\mathrm{s}}$ the vector of the electric potential output from piezoelectric material layers. The in-balance force and charge vectors, the external force and charge vectors are calculated by

$$ {}^{1}\boldsymbol{F}_{\mathrm{ui}} = {}^{1}\boldsymbol{F}_{\mathrm{uu}} + {}^{1}\boldsymbol{F}_{\mathrm{u}\phi}\,, \tag{5.69} $$

$$ {}^{1}\boldsymbol{G}_{\phi\mathrm{i}} = {}^{1}\boldsymbol{F}_{\phi\mathrm{u}} + {}^{1}\boldsymbol{F}_{\phi\phi}\,, \tag{5.70} $$

$$ \boldsymbol{F}_{\mathrm{ue}} = \boldsymbol{F}_{\mathrm{ub}} + \boldsymbol{F}_{\mathrm{us}} + \boldsymbol{F}_{\mathrm{uc}}\,, \tag{5.71} $$

$$ \boldsymbol{G}_{\phi\mathrm{e}} = \boldsymbol{G}_{\phi\mathrm{s}} + \boldsymbol{G}_{\phi\mathrm{c}}\,. \tag{5.72} $$

The dynamic equations derived by finite element method exclude damping matrix. Precise damping effect of a system is very difficult to model. However, for simulation purposes, the damping matrix can be calculated by linear summation of mass and stiffness matrices. The Rayleigh damping coefficients computation method [13] is an efficient way, which is given by

$$ {}^{1}\boldsymbol{C}_{\mathrm{uu}} = \frac{\alpha_1 + \alpha_2}{2}\,{}^{1}\boldsymbol{M}_{\mathrm{uu}} + \frac{\beta_1 + \beta_2}{2}\,{}^{1}\bar{\boldsymbol{K}}_{\mathrm{uu}}\,. \tag{5.73} $$

Here the coefficients α_1, α_2, β_1 and β_2 can be calculated as

$$ \begin{aligned} \beta_1 &= \frac{2(\varsigma_1\omega_1 - \varsigma_m\omega_m)}{\omega_1^2 - \omega_m^2}\,, & \alpha_1 &= 2\varsigma_1\omega_1 - \beta_1\omega_1^2\,, \\ \beta_2 &= \frac{2(\varsigma_1\omega_1 - \varsigma_{2.5m}\omega_{2.5m})}{\omega_1^2 - \omega_{2.5m}^2}\,, & \alpha_2 &= 2\varsigma_1\omega_1 - \beta_2\omega_1^2\,. \end{aligned} \tag{5.74} $$

In Eq. (5.74), ς_1, ς_m and $\varsigma_{2.5m}$ ($m = 2, 4, 6, \ldots$) refer to the damping ratio at 1, m and 2.5m modes, respectively. Similarly, ω_1, ω_m and $\omega_{2.5m}$ are the angular frequencies at 1, m and 2.5m modes. The damping ratio at ith mode can be assumed as

$$ \varsigma_i = \begin{cases} \dfrac{\varsigma_m - \varsigma_1}{\omega_m - \omega_1}(\omega_i - \omega_1) + \varsigma_1 & 1 < i < m \\[2ex] \dfrac{\varsigma_m - \varsigma_1}{\omega_m - \omega_1}(\omega_{m+i} - \omega_m) + \varsigma_1 & m < i < 2.5m\,. \end{cases} \tag{5.75} $$

Adding the damping coefficient matrix yields the equation of motion with considering the damping effects as

$$ {}^{1}\boldsymbol{M}_{\mathrm{uu}}{}_{0}^{2}\ddot{\boldsymbol{q}} + {}^{1}\boldsymbol{C}_{\mathrm{uu}}{}_{0}^{2}\dot{\boldsymbol{q}} + {}^{1}\boldsymbol{K}_{\mathrm{uu}}\Delta\boldsymbol{q} + {}^{1}\boldsymbol{K}_{\mathrm{u}\phi}\Delta\boldsymbol{\phi}_{a} = \boldsymbol{F}_{\mathrm{ue}} - {}^{1}\boldsymbol{F}_{\mathrm{ui}}\,, \tag{5.76} $$

where ${}_{0}^{2}\dot{\boldsymbol{q}}$ represents the velocity of the nodal DOF vector in the virtual configuration.

5.6.2 *Static FE Model*

By applying the FE method and the principle of virtual work, an electro-mechanically coupled static FE model including an equilibrium equation and a sensor equation for smart structures can be obtained as

$$ {}^{1}\bar{\boldsymbol{K}}_{\mathrm{uu}}\Delta\boldsymbol{q} + {}^{1}\boldsymbol{K}_{\mathrm{u}\phi}\Delta\boldsymbol{\phi}_{a} = \boldsymbol{F}_{\mathrm{ue}} - {}^{1}\boldsymbol{F}_{\mathrm{ui}}\,, \tag{5.77} $$

$$ {}^{1}\boldsymbol{K}_{\phi\mathrm{u}}\Delta\boldsymbol{q} + {}^{1}\boldsymbol{K}_{\phi\phi}\Delta\boldsymbol{\phi}_{s} = \boldsymbol{G}_{\phi\mathrm{e}} - {}^{1}\boldsymbol{G}_{\phi\mathrm{i}}\,. \tag{5.78} $$

Here, the coefficient matrices have the same meanings as those given in Sect. 5.6.1.

5.7 Geometrically and Electroelastic Nonlinear FE Model

Geometrically nonlinear phenomenon should be considered when smart structures undergo large displacements and rotations. At the meantime, electroelastic materially nonlinear effect should be included in the model when structures under strong electric field. Including both geometrically nonlinear and materially nonlinear phenomena, electroelastic nonlinear constitutive equations given in Eq. (4.48) and Green-Lagrange nonlinear strains given in Eqs. (3.83)–(3.88) have to be employed in the mathematical model. The internal virtual work that considers both geometrically and materially nonlinear is given as

$$ \begin{aligned} \delta W_{\mathrm{int}} = \int_{\mathrm{V}} \Bigg(& \delta\boldsymbol{\varepsilon}^{\mathrm{T}}\left(\boldsymbol{c}^{\mathrm{E}}\boldsymbol{\varepsilon} - \boldsymbol{e}^{\mathrm{T}}\boldsymbol{E} - \frac{1}{2}\boldsymbol{b}|\bar{\boldsymbol{E}}|\boldsymbol{E}\right) \\ & - \delta\boldsymbol{E}^{\mathrm{T}}\left(\boldsymbol{e}\boldsymbol{\varepsilon} + \boldsymbol{g}^{\mathrm{S}}\boldsymbol{E} + \frac{1}{2}\boldsymbol{h}|\bar{\boldsymbol{E}}|\boldsymbol{E}\right)\Bigg)\,\mathrm{dV}\,. \end{aligned} \tag{5.79} $$

Using the resultant strain vectors and taking the integral of the thickness coordinate the internal virtual work can be expressed as

$$\delta W_{\text{int}} = \int_{V} \Big(\delta \boldsymbol{S}^{\mathrm{T}} \big(\boldsymbol{H}_{\mathrm{c}} \boldsymbol{S} + \boldsymbol{H}_{\mathrm{e}}^{\mathrm{T}} \boldsymbol{E} + \boldsymbol{H}_{\mathrm{b}} |\bar{\boldsymbol{E}}| \boldsymbol{E} \big) + \delta \boldsymbol{E}^{\mathrm{T}} \big(\boldsymbol{H}_{\mathrm{e}} \boldsymbol{S} + \boldsymbol{H}_{\mathrm{g}} \boldsymbol{E} + \boldsymbol{H}_{\mathrm{h}} |\bar{\boldsymbol{E}}| \boldsymbol{E} \big) \Big) \, \mathrm{dV} \,. \tag{5.80}$$

Here the matrices $\boldsymbol{H}_{\mathrm{c}}$, $\boldsymbol{H}_{\mathrm{e}}$ and $\boldsymbol{H}_{\mathrm{g}}$ have already been discussed in Eqs. (5.38)–(5.40). In a similar way, the matrices $\boldsymbol{H}_{\mathrm{b}}$ and $\boldsymbol{H}_{\mathrm{h}}$ can be obtained as

$$\boldsymbol{H}_{\mathrm{b}} = -\int_{\mathrm{h}} \frac{1}{2} \boldsymbol{H}_{\mathrm{s}}^{\mathrm{T}} \, \boldsymbol{b} \, \mu \, \mathrm{d}\Theta^3 \,, \tag{5.81}$$

$$\boldsymbol{H}_{\mathrm{h}} = -\int_{\mathrm{h}} \frac{1}{2} \boldsymbol{h} \, \mu \, \mathrm{d}\Theta^3 \,. \tag{5.82}$$

Using the TL incremental formulation, the nonlinear electric filed coefficient matrix $\bar{\boldsymbol{E}}$ at configuration ${}^2\mathrm{C}$ can be expressed by a summation of those at configuration ${}^1\mathrm{C}$ and the incremental vector as

$${}^2_0|\bar{\boldsymbol{E}}| = {}^1_0|\bar{\boldsymbol{E}}| + |\Delta \bar{\boldsymbol{E}}|. \tag{5.83}$$

Then the term ${}^2_0|\bar{\boldsymbol{E}}| \, {}^2_0\boldsymbol{E}$ can be obtained as

$$\begin{aligned} {}^2_0|\bar{\boldsymbol{E}}| \, {}^2_0\boldsymbol{E} &= \big({}^1_0|\bar{\boldsymbol{E}}| + |\Delta \bar{\boldsymbol{E}}| \big) \big({}^1_0\bar{\boldsymbol{E}} + \Delta \bar{\boldsymbol{E}} \big) \\ &= {}^1_0|\bar{\boldsymbol{E}}| \, {}^1_0\bar{\boldsymbol{E}} + 2 \, {}^1_0|\bar{\boldsymbol{E}}| |\Delta \bar{\boldsymbol{E}}| + |\Delta \bar{\boldsymbol{E}}| \, \Delta \bar{\boldsymbol{E}} \,. \end{aligned} \tag{5.84}$$

Using the above equations, the internal virtual work at configuration ${}^2\mathrm{C}$ can be re-arranged as

$$\begin{aligned} {}^2_0\delta W_{\text{int}} &= \int_{\Omega} \big({}^2_0\delta\boldsymbol{\varepsilon}^{\mathrm{T}} \, {}^2_0\boldsymbol{\sigma} - {}^2_0\delta\boldsymbol{E}^{\mathrm{T}} \, {}^2_0\boldsymbol{D} \big) \, \mathrm{d}\Omega \\ &= \delta\Delta\boldsymbol{q}^{\mathrm{T}} \int_{\Omega} \Big(\boldsymbol{B}_{\mathrm{u}}^{\mathrm{T}} \boldsymbol{H}_{\mathrm{c}} \, {}^1_0\boldsymbol{S} + \boldsymbol{B}_{\mathrm{u}}^{\mathrm{T}} \boldsymbol{H}_{\mathrm{c}} \, \boldsymbol{B}_{\mathrm{u}} \Delta\boldsymbol{q} + \boldsymbol{B}_{\mathrm{u}}^{\mathrm{T}} \boldsymbol{H}_{\mathrm{e}}^{\mathrm{T}} \, {}^1_0\boldsymbol{E} + \boldsymbol{B}_{\mathrm{u}}^{\mathrm{T}} \boldsymbol{H}_{\mathrm{e}}^{\mathrm{T}} \, \boldsymbol{B}_{\phi} \Delta\boldsymbol{\phi} \\ &\quad + \boldsymbol{B}_{\mathrm{u}}^{\mathrm{T}} \boldsymbol{H}_{\mathrm{b}}^{\mathrm{T}} \, {}^1_0|\bar{\boldsymbol{E}}| \, {}^1_0\boldsymbol{E} + 2\boldsymbol{B}_{\mathrm{u}}^{\mathrm{T}} \boldsymbol{H}_{\mathrm{b}}^{\mathrm{T}} \, {}^1_0|\bar{\boldsymbol{E}}| \boldsymbol{B}_{\phi} \Delta\boldsymbol{\phi} + \underline{\boldsymbol{B}_{\mathrm{u}}^{\mathrm{T}} \boldsymbol{H}_{\mathrm{b}}^{\mathrm{T}} |\Delta\bar{\boldsymbol{E}}| \Delta\boldsymbol{E}} \Big) \mathrm{d}\Omega \\ &\quad + \delta\Delta\boldsymbol{\phi}^{\mathrm{T}} \int_{\Omega} \Big(\boldsymbol{B}_{\phi}^{\mathrm{T}} \boldsymbol{H}_{\mathrm{e}} \, {}^1_0\boldsymbol{S} + \boldsymbol{B}_{\phi}^{\mathrm{T}} \boldsymbol{H}_{\mathrm{e}} \, \boldsymbol{B}_{\mathrm{u}} \Delta\boldsymbol{q} + \boldsymbol{B}_{\phi}^{\mathrm{T}} \boldsymbol{H}_{\mathrm{g}} \, {}^1_0\boldsymbol{E} + \boldsymbol{B}_{\phi}^{\mathrm{T}} \boldsymbol{H}_{\mathrm{g}} \, \boldsymbol{B}_{\phi} \Delta\boldsymbol{\phi} \\ &\quad + \boldsymbol{B}_{\phi}^{\mathrm{T}} \boldsymbol{H}_{\mathrm{h}} \, {}^1_0|\bar{\boldsymbol{E}}| \, {}^1_0\boldsymbol{E} + 2\boldsymbol{B}_{\phi}^{\mathrm{T}} \boldsymbol{H}_{\mathrm{h}} \, {}^1_0|\bar{\boldsymbol{E}}| \boldsymbol{B}_{\phi} \Delta\boldsymbol{\phi} + \underline{\boldsymbol{B}_{\phi}^{\mathrm{T}} \boldsymbol{H}_{\mathrm{h}} |\Delta\bar{\boldsymbol{E}}| \Delta\boldsymbol{E}} \Big) \, \mathrm{d}\Omega \\ &= \delta\Delta\boldsymbol{q}^{\mathrm{T}} \big({}^1\boldsymbol{K}_{\mathrm{uu}} \Delta\boldsymbol{q} + {}^1\boldsymbol{K}_{\mathrm{u}\phi} \Delta\boldsymbol{\phi} + {}^1\boldsymbol{F}_{\mathrm{ui}} \big) \\ &\quad + \delta\Delta\boldsymbol{\phi}^{\mathrm{T}} \big({}^1\boldsymbol{K}_{\phi\mathrm{u}} \Delta\boldsymbol{q} + {}^1\boldsymbol{K}_{\phi\phi} \Delta\boldsymbol{\phi} + {}^1\boldsymbol{G}_{\phi\mathrm{i}} \big) \,, \end{aligned} \tag{5.85}$$

where

$$ {}^{1}\boldsymbol{K}_{\mathrm{uu}} = \int_{\Omega} \boldsymbol{B}_{\mathrm{u}}^{\mathrm{T}} \boldsymbol{H}_{\mathrm{c}}\, \boldsymbol{B}_{\mathrm{u}}\, \mathrm{d}\Omega, \tag{5.86} $$

$$ {}^{1}\boldsymbol{K}_{\mathrm{u}\phi} = \int_{\Omega} (\boldsymbol{B}_{\mathrm{u}}^{\mathrm{T}} \boldsymbol{H}_{\mathrm{e}}^{\mathrm{T}}\, \boldsymbol{B}_{\phi} + 2\boldsymbol{B}_{\mathrm{u}}^{\mathrm{T}} \boldsymbol{H}_{\mathrm{b}}^{\mathrm{T}}\, {}^{1}_{0}|\bar{\boldsymbol{E}}| \boldsymbol{B}_{\phi})\, \mathrm{d}\Omega, \tag{5.87} $$

$$ {}^{1}\boldsymbol{F}_{\mathrm{ui}} = \int_{\Omega} (\boldsymbol{B}_{\mathrm{u}}^{\mathrm{T}} \boldsymbol{H}_{\mathrm{c}}\, {}^{1}_{0}\boldsymbol{S} + \boldsymbol{B}_{\mathrm{u}}^{\mathrm{T}} \boldsymbol{H}_{\mathrm{e}}^{\mathrm{T}}\, {}^{1}_{0}\boldsymbol{E} + \boldsymbol{B}_{\mathrm{u}}^{\mathrm{T}} \boldsymbol{H}_{\mathrm{b}}^{\mathrm{T}}\, {}^{1}_{0}|\bar{\boldsymbol{E}}|\, {}^{1}_{0}\boldsymbol{E})\, \mathrm{d}\Omega, \tag{5.88} $$

$$ {}^{1}\boldsymbol{K}_{\phi\mathrm{u}} = \int_{\Omega} \boldsymbol{B}_{\phi}^{\mathrm{T}} \boldsymbol{H}_{\mathrm{e}}\, \boldsymbol{B}_{\mathrm{u}}\, \mathrm{d}\Omega, \tag{5.89} $$

$$ {}^{1}\boldsymbol{K}_{\phi\phi} = \int_{\Omega} (\boldsymbol{B}_{\phi}^{\mathrm{T}} \boldsymbol{H}_{\mathrm{g}}\, \boldsymbol{B}_{\phi} + 2\boldsymbol{B}_{\phi}^{\mathrm{T}} \boldsymbol{H}_{\mathrm{h}}\, {}^{1}_{0}|\bar{\boldsymbol{E}}| \boldsymbol{B}_{\phi})\, \mathrm{d}\Omega, \tag{5.90} $$

$$ {}^{1}\boldsymbol{G}_{\phi\mathrm{i}} = \int_{\Omega} (\boldsymbol{B}_{\phi}^{\mathrm{T}} \boldsymbol{H}_{\mathrm{e}}\, {}^{1}_{0}\boldsymbol{S} + \boldsymbol{B}_{\phi}^{\mathrm{T}} \boldsymbol{H}_{\mathrm{g}}\, {}^{1}_{0}\boldsymbol{E} + \boldsymbol{B}_{\phi}^{\mathrm{T}} \boldsymbol{H}_{\mathrm{h}}\, {}^{1}_{0}|\bar{\boldsymbol{E}}|\, {}^{1}_{0}\boldsymbol{E})\, \mathrm{d}\Omega. \tag{5.91} $$

Additionally, the terms underlined are neglected due to the second order of the infinitesimal electric field increment. In this case, the coupled coefficient matrices ${}^{1}\boldsymbol{K}_{\mathrm{u}\phi}$ and ${}^{1}\boldsymbol{K}_{\phi\mathrm{u}}$ are no long symmetric to each other. For more details, we refer to our publications Ref. [14].

5.8 Numerical Algorithms

The equations of motion and the equilibrium equations of smart structures have been constructed using finite element method. The equations of motion are second-order differential equations with respect to time. The Newmark method (implicit method) and the Central Difference Algorithm (CDA, explicit method) are employed for solving the second-order differential equations of motion. Between these two methods, Newmark method is used much more frequently in dynamic analysis than CDA, because of high computational efficiency and robustness.

For the static equilibrium equations, zero-order differential equations, load control method and arc-length control method are the most used ones. Newton-Raphson method is the load control method, which can be used to calculate monotonic nonlinear static response. The arc-length control method, Riks-Wempner method, is used for buckling and post-buckling analysis. The details of these numerical algorithms can be found in many books or thesis, see e.g. [4, 8, 10–12] among many others.

5.8.1 Newmark Method

The dynamic equation at time $t + \Delta t$ is considered as

$$ \boldsymbol{M}_{\mathrm{uu}}^{(t)} \ddot{\boldsymbol{q}}^{(t+\Delta t)} + \boldsymbol{C}_{\mathrm{uu}}^{(t)} \dot{\boldsymbol{q}}^{(t+\Delta t)} + \boldsymbol{K}_{\mathrm{uu}}^{(t)} \Delta \boldsymbol{q}^{(t)} = \boldsymbol{F}_{\mathrm{ue}}^{(t)} - \boldsymbol{F}_{\mathrm{ui}}^{(t)} \tag{5.92} $$

with the assumptions of $\boldsymbol{q}$ and $\dot{\boldsymbol{q}}$ at time $t + \Delta t$ as [10, 11]

$$\boldsymbol{q}^{(t+\Delta t)} = \boldsymbol{q}^{(t)} + (\Delta t)\dot{\boldsymbol{q}}^{(t)} + (\Delta t)^2\big((0.5-\beta)\ddot{\boldsymbol{q}}^{(t)} + \beta\ddot{\boldsymbol{q}}^{(t+\Delta t)}\big)\,, \tag{5.93}$$

$$\dot{\boldsymbol{q}}^{(t+\Delta t)} = \dot{\boldsymbol{q}}^{(t)} + (\Delta t)\big((1-\gamma)\ddot{\boldsymbol{q}}^{(t)} + \gamma\ddot{\boldsymbol{q}}^{(t+\Delta t)}\big)\,. \tag{5.94}$$

Here superscript t refers to the time in the current configuration, and Δt is a small increment of time.

If the parameters satisfy $\gamma \geqslant 0.5$ and $\beta \geqslant (2\gamma+1)^2/16$, the Newmark method is unconditionally stable [6], meaning that the length of time step has no effect on the stability of the solution, but it has influence on the accuracy. The commonly used values are $\beta = 0.25$ and $\gamma = 0.5$, with which it is called linear acceleration method. For simplicity, some constants will be introduced for calculation as

$$\begin{aligned} a_0 &= \frac{1}{\beta(\Delta t)^2}\,, \quad a_1 = \frac{\gamma}{\beta(\Delta t)}\,, \quad a_2 = \frac{1}{\beta(\Delta t)}\,, \quad a_3 = \frac{\gamma}{\beta}\,, \\ a_4 &= \frac{1}{2\beta}\,, \quad a_5 = \Big(1 - \frac{\gamma}{2\beta}\Big)(\Delta t)\,, \quad a_6 = 1 - \frac{1}{2\beta}\,, \quad a_7 = 1 - \frac{\gamma}{\beta}\,. \end{aligned} \tag{5.95}$$

Using the assumptions in (5.93) and (5.94), the incremental acceleration and velocity of the nodal displacement vector can be obtained as

$$\Delta\ddot{\boldsymbol{q}}^{(t)} = a_0\Delta\boldsymbol{q}^{(t)} - a_2\dot{\boldsymbol{q}}^{(t)} - a_4\ddot{\boldsymbol{q}}^{(t)}\,, \tag{5.96}$$

$$\Delta\dot{\boldsymbol{q}}^{(t)} = a_1\Delta\boldsymbol{q}^{(t)} - a_3\dot{\boldsymbol{q}}^{(t)} + a_5\ddot{\boldsymbol{q}}^{(t)}\,. \tag{5.97}$$

Substituting Eqs. (5.96) and (5.97) into (5.92) yields

$$\Delta\boldsymbol{q}^{(t)} = \frac{\boldsymbol{F}_{\mathrm{ue}}^{(t)} - \boldsymbol{F}_{\mathrm{ui}}^{(t)} - \big(a_6\boldsymbol{M}_{\mathrm{uu}}^{(t)} + a_5\boldsymbol{C}_{\mathrm{uu}}^{(t)}\big)\ddot{\boldsymbol{q}}^{(t)} - \big(a_7\boldsymbol{C}_{\mathrm{uu}}^{(t)} - a_2\boldsymbol{M}_{\mathrm{uu}}^{(t)}\big)\dot{\boldsymbol{q}}^{(t)}}{a_0\boldsymbol{M}_{\mathrm{uu}}^{(t)} + a_1\boldsymbol{C}_{\mathrm{uu}}^{(t)} + \boldsymbol{K}_{\mathrm{uu}}^{(t)}}\,. \tag{5.98}$$

5.8.2 Central Difference Algorithm

Regarding to the central difference algorithm, the equations of motion at time t are considered as

$$\boldsymbol{M}_{\mathrm{uu}}^{(t)}\ddot{\boldsymbol{q}}^{(t)} + \boldsymbol{C}_{\mathrm{uu}}^{(t)}\dot{\boldsymbol{q}}^{(t)} + \boldsymbol{K}_{\mathrm{uu}}^{(t)}\Delta\boldsymbol{q}^{(t)} = \boldsymbol{F}_{\mathrm{ue}}^{(t)} - \boldsymbol{F}_{\mathrm{ui}}^{(t)}\,, \tag{5.99}$$

with the approximations of acceleration $\ddot{\boldsymbol{q}}^{(t)}$ and velocity $\dot{\boldsymbol{q}}^{(t)}$ at time t as [10, 11]

$$\ddot{\boldsymbol{q}}^{(t)} = \frac{1}{(\Delta t)^2}\Big(\boldsymbol{q}^{(t+\Delta t)} - 2\boldsymbol{q}^{(t)} + \boldsymbol{q}^{(t-\Delta t)}\Big)\,, \tag{5.100}$$

$$\dot{\boldsymbol{q}}^{(t)} = \frac{1}{2(\Delta t)}\Big(\boldsymbol{q}^{(t+\Delta t)} - \boldsymbol{q}^{(t-\Delta t)}\Big)\,. \tag{5.101}$$

Substituting the above assumptions into the dynamic equation one obtains the displacement vector at time $t + \Delta t$ as

$$\boldsymbol{q}^{(t+\Delta t)} = \left(\frac{1}{(\Delta t)^2}\boldsymbol{M}_{\mathrm{uu}}^{(t)} + \frac{1}{2(\Delta t)}\boldsymbol{C}_{\mathrm{uu}}^{(t)}\right)^{-1}\boldsymbol{F}_{\mathrm{Residual}}^{(t)} \tag{5.102}$$

where

$$\begin{aligned}\boldsymbol{F}_{\mathrm{Residual}}^{(t)} = {} & \boldsymbol{F}_{\mathrm{ue}}^{(t)} - \boldsymbol{F}_{\mathrm{ui}}^{(t)} + \frac{1}{(\Delta t)^2}\boldsymbol{M}_{\mathrm{uu}}^{(t)}\big(2\boldsymbol{q}^{(t)} - \boldsymbol{q}^{(t-\Delta t)}\big) \\ & + \frac{1}{2(\Delta t)}\boldsymbol{C}_{\mathrm{uu}}^{(t)}\boldsymbol{q}^{(t-\Delta t)} + \boldsymbol{K}_{\mathrm{uu}}^{(t)}\big(\boldsymbol{q}^{(t-\Delta t)} - \boldsymbol{q}^{(t)}\big)\,.\end{aligned} \tag{5.103}$$

For the first step of CDA, the displacement at time $t - \Delta t$ is needed, which can be built as

$$\boldsymbol{q}^{(t-\Delta t)} = \boldsymbol{q}^{(t)} + (\Delta t)\dot{\boldsymbol{q}}^{(t)} + \frac{(\Delta t)^2}{2}\ddot{\boldsymbol{q}}^{(t)}\,, \tag{5.104}$$

where $\boldsymbol{q}^{(t)}$ and $\dot{\boldsymbol{q}}^{(t)}$ are prescribed, and $\ddot{\boldsymbol{q}}^{(t)}$ can be determined by

$$\ddot{\boldsymbol{q}}^{(t)} = (\boldsymbol{M}_{\mathrm{uu}}^{(t)})^{-1}\Big(\boldsymbol{F}_{\mathrm{ue}}^{(t)} - \boldsymbol{F}_{\mathrm{ui}}^{(t)} - \boldsymbol{C}_{\mathrm{uu}}^{(t)}\dot{\boldsymbol{q}}^{(t)} - \boldsymbol{K}_{\mathrm{uu}}^{(t)}\Delta\boldsymbol{q}^{(t)}\Big)\,. \tag{5.105}$$

5.8.3 Newton-Raphson Method

For nonlinear static equilibrium equation, the calculation iteration is investigated. The kth iteration of static equilibrium equation is defined as

$$\boldsymbol{K}_{\mathrm{uu}}^{(k)}\Delta\boldsymbol{q}^{(k)} = \boldsymbol{F}_{\mathrm{ue}}^{(k)} - \boldsymbol{F}_{\mathrm{ui}}^{(k)}\,. \tag{5.106}$$

Thus, the incremental displacement vector in the kth iteration can be solved by

$$\Delta\boldsymbol{q}^{(k)} = (\boldsymbol{K}_{\mathrm{uu}}^{(k)})^{-1}\Big(\boldsymbol{F}_{\mathrm{ue}}^{(k)} - \boldsymbol{F}_{\mathrm{ui}}^{(k)}\Big)\,. \tag{5.107}$$

Consequently, the displacement vector in iteration $k + 1$ can be obtained as

$$\boldsymbol{q}^{(k+1)} = \boldsymbol{q}^{(k)} + \Delta\boldsymbol{q}^{(k)}\,. \tag{5.108}$$

Using the current displacement matrix, the system matrices and vectors will be updated to $\boldsymbol{K}_{\mathrm{uu}}^{(k+1)}$, $\boldsymbol{F}_{\mathrm{ue}}^{(k+1)}$ and $\boldsymbol{F}_{\mathrm{ui}}^{(k+1)}$. In such a way, the equilibrium equation changes to the $(k + 1)$th iteration. Again, $\Delta\boldsymbol{q}$ can be calculated, and new equilibrium equation is formed. Following the above loops, until it converges within an accepted error ϵ as

$$\frac{\|\Delta \boldsymbol{q}^{(k)}\|}{\|\boldsymbol{q}^{(k+1)}\|} < \epsilon\,. \tag{5.109}$$

5.8.4 Riks-Wempner Method

The Riks-Wempner algorithm is an arc-length control method, used for solving nonlinear equilibrium equations, especially for buckling analysis. The algorithm can be found in many books e.g. [8, 15]. The strategies of searching the equilibrium point is critical important. Two approaches for determine the searching direction, namely along the normal plane or the spherical surface. In this report, the iteration procedure of the Riks-Wempner algorithm goes along the normal plane with stiffness matrices updated in every iteration, as shown in Fig. 5.6.

Introducing a proportional loading factor, the nonlinear equilibrium equation in the ith iteration can be re-written as

$$^{1}\bar{\boldsymbol{K}}_{\mathrm{uu}}^{(i)}\Delta \boldsymbol{q}^{(i)} = \lambda^{(i)}\boldsymbol{F}_{\mathrm{ue}}^{(i)} - {}^{1}\boldsymbol{F}_{\mathrm{ui}}^{(i)}\,. \tag{5.110}$$

Here, $\lambda^{(i)}$ denotes the proportional loading factor, varying between 0 and 1. The incremental displacement vector $\Delta \boldsymbol{q}^{(i)}$ can be calculated by the linearized equilib-

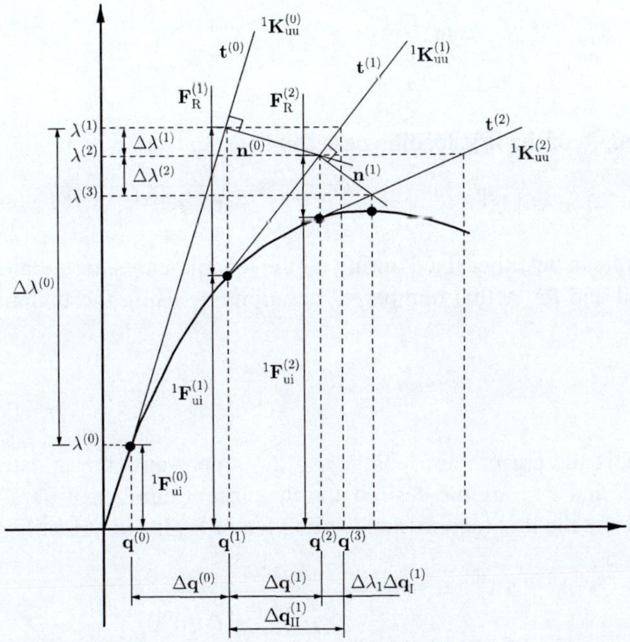

Fig. 5.6 Schematic procedure for the Riks-Wempner method

rium equation as

$$^{1}\bar{\boldsymbol{K}}_{\mathrm{uu}}^{(i)} \Delta \boldsymbol{q}^{(i)} = \Delta\lambda^{(i)} \boldsymbol{F}_{\mathrm{ue}}^{(i)} . \tag{5.111}$$

Further, we define a tangent vector of the equilibrium path at 0th and ith iteration as

$$\boldsymbol{t}^{(0)} = \begin{Bmatrix} \Delta\boldsymbol{q}^{(0)} \\ \Delta\lambda^{(0)} \end{Bmatrix} , \quad \boldsymbol{t}^{(i)} = \begin{Bmatrix} \Delta\lambda^{(i)} \Delta{\boldsymbol{q}_{\mathrm{I}}}^{(i)} \\ \Delta\lambda^{(i)} \end{Bmatrix} \quad (i \geqslant 1) . \tag{5.112}$$

The searching orientation vector $\boldsymbol{n}^{(i)}$, normal to the tangent vector $\boldsymbol{t}^{(i)}$, can be defined as

$$\boldsymbol{n}^{(i)} = \begin{Bmatrix} \Delta\boldsymbol{q}^{(i+1)} \\ -\Delta\lambda^{(i+1)} \end{Bmatrix} . \tag{5.113}$$

The initial increment of the loading factor $\Delta\lambda^{(0)}$ is prescribed, and the next incremental loading factor can be obtained by using the constraints of $\boldsymbol{n}^{(i)} \cdot \boldsymbol{t}^{(i)} = 0$ as

$$\Delta\lambda^{(i)} = \begin{cases} \dfrac{(\Delta\boldsymbol{q}^{(0)})^{\mathrm{T}} \Delta{\boldsymbol{q}_{\mathrm{II}}}^{(1)}}{(\Delta\boldsymbol{q}^{(0)})^{\mathrm{T}} \Delta{\boldsymbol{q}_{\mathrm{I}}}^{(1)} + \Delta\lambda^{(0)}} & i = 1 \\[2ex] \dfrac{(\Delta{\boldsymbol{q}_{\mathrm{I}}}^{(i-1)})^{\mathrm{T}} \Delta{\boldsymbol{q}_{\mathrm{II}}}^{(i)}}{(\Delta{\boldsymbol{q}_{\mathrm{I}}}^{(i-1)})^{\mathrm{T}} \Delta{\boldsymbol{q}_{\mathrm{I}}}^{(i)} + 1} & i \geqslant 2 \end{cases} \tag{5.114}$$

with

$$\Delta{\boldsymbol{q}_{\mathrm{I}}}^{(i)} = ({}^{1}\boldsymbol{K}_{\mathrm{uu}}^{(i)})^{-1} \boldsymbol{F}_{\mathrm{ue}}^{(i)} , s\Delta{\boldsymbol{q}_{\mathrm{II}}}^{(i)} \quad = ({}^{1}\boldsymbol{K}_{\mathrm{uu}}^{(i)})^{-1} \left({}^{1}\boldsymbol{K}_{\mathrm{uu}}^{(i-1)} \Delta\boldsymbol{q}^{(i-1)} + {}^{1}\boldsymbol{F}_{\mathrm{ui}}^{(i-1)} - {}^{1}\boldsymbol{F}_{\mathrm{ui}}^{(i)} \right) . \tag{5.115}$$

The arc length of the first loading case can be calculated by

$$\Delta S_0 = \|\boldsymbol{t}^{(0)} \cdot \boldsymbol{t}^{(0)}\| = \sqrt{(\Delta\lambda^{(0)})^2 + (\Delta\boldsymbol{q}^{(0)})^{\mathrm{T}} \cdot \Delta\boldsymbol{q}^{(0)}} . \tag{5.116}$$

The arc length can be either fixed during all the loading cases, or updated according to the desired and the actual number of iterations by using the updating equation given as

$$\Delta S_i = \Delta S_{i-1} \sqrt{\frac{I_{des}}{I_{i-1}}} , \tag{5.117}$$

where ΔS_{i-1} is the current arc length, and ΔS_i represents the updated one. Furthermore, I_{des} and I_{i-1} are the desired and the current number of iteration, respectively. Therefore, the first incremental loading factor for the next loading step can be obtained as

$$\Delta\lambda_i^{(0)} = \frac{\pm\Delta S_i}{\sqrt{1 + (\Delta{\boldsymbol{q}_{\mathrm{I}}}^{(0)})_i^{\mathrm{T}} \cdot (\Delta{\boldsymbol{q}_{\mathrm{I}}}^{(0)})_i}} . \tag{5.118}$$

Here, the sign of ΔS_i can be determined by the stiffness matrix.

5.9 Summary

This chapter developed static and dynamic nonlinear FE models for piezoelectric integrated smart structures undergoing large displacements and under strong electric fields. In the FE models, both linear piezoelectric constitutive equations and electroelastic nonlinear constitutive equations were considered, where electric field through the thickness are assumed constant. Four types of shell elements were developed for composite or piezoelectric laminated thin-walled structures. Various geometrically nonlinear strain-displacement relations were included in the FE models. In LRT56 FE models, Euler angles were used to represent unrestricted finite rotations in shell structures. In the end of this chapter, several numerical algorithms for solving static and dynamic equations were discussed.

References

1. L. Librescu, *Elastostatics and Kinetics of Anisotropic and Heterogeneous Shell-Type Structures* (Noordhoff International, Leyden, 1975)
2. S.Q. Zhang, R. Schmidt, Large rotation theory for static analysis of composite and piezoelectric laminated thin-walled structures. Thin-Walled Struct. **78**, 16–25 (2014)
3. S.Q. Zhang, R. Schmidt, Large rotation FE transient analysis of piezolaminated thin-walled smart structures. Smart Mater. Struct. **22**, 105025 (2013)
4. I. Kreja, R. Schmidt, Large rotations in first-order shear deformation FE analysis of laminated shells. Int. J. Non-Linear Mech. **41**, 101–123 (2006)
5. G.A. Korn, T.M. Korn, *Mathematical Handbook for Scientists and Engineers* (Dover Publication Inc., Mineola, New York, 2000)
6. K.J. Bathe, *Finite Element Procedures* (Prentice Hall, New Jersey, 1996)
7. O.C. Zienkiewicz, R.L. Taylor, J.Z. Zhu, *The Finite Element Method: Its Basis and Fundamentals*, 6th edn. (Butterworth-Heinemann, Oxford, 2005)
8. I. Kreja, *Geometrically non-linear analysis of layered composite plates and shells*. Habilitation Thesis, Published as Monografie 83, Politechnika Gdańska (2007)
9. V. Piefort, *Finite element modeling of piezoelectric active structures*. Ph.D. Thesis, Universite Libre de Bruxelles (2001)
10. D.T. Vu, *Geometrically nonlinear higher-oder shear deformation FE analysis of thin-walled smart structures*. Ph.D. Thesis, RWTH Aachen University (2011)
11. S. Lentzen, *Nonlinear coupled thermopiezoelectric modelling and FE-simulation of smart structures*. Ph.D. Thesis, RWTH Aachen University, in: Fortschritt-Berichte VDI, Reihe 20, Nr. 419, VDI Verlag, Düsseldorf (2009)
12. M.A. Crisfield, *Non-linear Finite Element Analysis of Solids and Structures* (Wiley, 1991)
13. I. Chowdhury, S.P. Dasgupta, Computation of Rayleigh damping coefficients for large systems. Electron. J. Geotech. Eng. **8**, 10392376 (2003)
14. S.Q. Zhang, G.Z. Zhao, S.Y. Zhang, R. Schmidt, X.S. Qin, Geometrically nonlinear FE analysis of piezoelectric laminated composite structures under strong driving electric field. Compos. Struct. **181**, 112–120 (2017)
15. C.L. Liao, J.N. Reddy, *An Incremental Total Lagrangian Formulation for General Anisotropic Shell-Type Structures* (Virginia Polytechnic Institute and State University, Blacksburg, Virginia, 1987)

Chapter 6
Nonlinear Analysis of Piezoceramic Laminated Structures

Abstract In this chapter, geometrically nonlinear FE models, including RVK5, MRT5, LRT5 and LRT56, are validated by static analysis of composite laminated thin-walled structures. Later, the geometrically nonlinear FE models are test by buckling and post-buckling analysis of cylindrical composite panels. Afterwards, the geometrically nonlinear FE models are implemented into static and dynamic analysis of piezolaminated beam, plate and shell structures. In the last part, the simulations of nonlinear phenomena including both geometrically and electroelastic materially nonlinear effects are performed through piezo structures undergoing large displacement and under strong electric field.

6.1 Benchmark Problems

6.1.1 Asymmetric Cross-Ply Laminated Plate

The first benchmark problem is an asymmetric cross-ply laminated plate, shown in Fig. 6.1, proposed by Sun et al. [1], and studied later by Reddy [2], Basar et al. [3], Kreja and Schmidt [4]. The composite plate consists of two substrate layers with a stacking sequence of $[90°/0°]$. The material angle 0° represents the fiber reinforcement being along Θ^1-axis, while material angle 90° denotes that being along Θ^2-axis. The in-plane dimensions are 9×15 in^2, and the thickness for each substrate layer is 0.02 in. The material properties are given in Table 6.1. The composite structure is meshed by 9×2 quadrilateral shell elements along the Θ^1- and Θ^2-axis, respectively. The boundary conditions of hinged and simply supported are considered. In the hinged case, all the DOFs are constrained except rotations around the Θ^2-axis. The simply supported boundary condition frees additionally the translational movement along the Θ^1-direction based on the hinged case. A surface force is applied on the plane with positive or negative, which is measured in the unit lb/in^2. The non-dimensional displacements $(|w|/h)$ of mid-point is calculated

For the hinged boundary condition, the mid-point displacements of the plate under two loading cases $\pm \boldsymbol{q}$ are derived using LRT56 theory, presented in Fig. 6.2a for small

S.-Q. Zhang, *Nonlinear Analysis of Thin-Walled Smart Structures*, Springer Tracts in Mechanical Engineering, https://doi.org/10.1007/978-981-15-9857-9_6

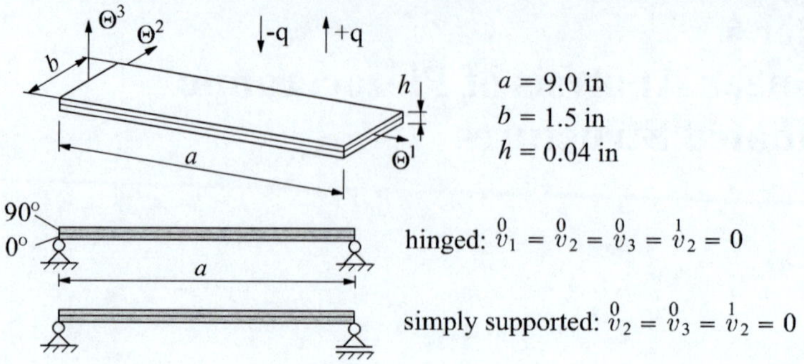

Fig. 6.1 Asymmetric cross-ply laminated plate

Table 6.1 Material properties of the composite plate

Orthotropic material
$E_1 = 2.0 \times 10^7$ lb/in^2
$E_2 = 1.4 \times 10^6$ lb/in^2
$\nu_{12} = \nu_{23} = 0.3$
$G_{12} = G_{23} = G_{13} = 0.7 \times 10^7$ lb/in^2

loading and (b) for large loading. The figures show that the present model predicts the displacements in excellent agreement with those predicted by the TOSD RVK5 theory of Reddy [2]. This is because that the plate of hinged boundary condition only deforms with small or moderate rotations. Structures with hinged boundary condition is difficult to deform with large rotations. This is also indicated by the results of Basar et al. [3], where TOSD large rotation theory was applied. The load-displacement curve shows complex path for the $+\boldsymbol{q}$ loading case, in which the structure first behaves softening then turning into a stress stiffening due to the nonlinear effect (Fig. 6.2a).

For the simply supported boundary condition under the same loading cases, the mid-point displacements are calculated and compared in Fig. 6.3, with the numerical values listed in Table 6.2. The results indicate that the present results obtained by LRT56 theory agree quite well with those presented in [3, 4]. The reference [4] developed the large rotation nonlinear model based on the FOSD hypothesis, while the reference [3] employed TOSD hypothesis.

Regarding to the results of RVK5, MRT5 and LRT5, large differences exist compared to those of LRT56. This is because the plate of simply supported boundary condition will deform in large rotations. The reasons of discrepancies between each simplified nonlinear models, RVK5, MRT5 and LRT5, are that different nonlinear strain-displacement relations are considered in each nonlinear model. The RVK5 theory only contains the squares and products of derivatives of the transverse deflection in the in-plane longitudinal and shear strain components. The MRT5 theory considers more strain-displacement nonlinear terms than RVK5, but with simplified relations compared with fully geometrically nonlinear terms. Both RVK5 and MRT5

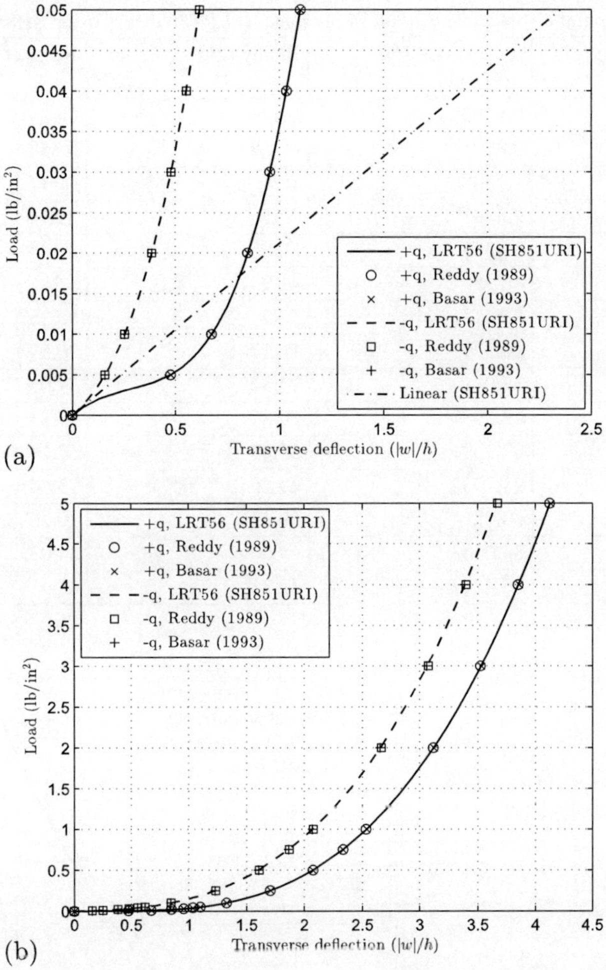

Fig. 6.2 Load-displacement curves of hinged cross-ply plate under a uniform pressure: **a** small pressure, **b** large pressure, reprinted from Ref. [5], copyright 2014, with permission from ELSEVIER

are under the assumption of small or moderate rotations. The LRT5 theory contains all the nonlinear strain-displacement relations. Since moderate rotations are assumed in the theory, the sixth parameter is neglected $\overset{1}{v}_3 = 0$, that means no updating of the rotations. Therefore, LRT5 theory cannot predict results more accurate than MRT5 theory.

Investigating on full integration and reduced integration elements, SH851FI and SH851URI, there are big discrepancies between the results of LRT56 using the two elements. The reason can be explained that the full integration converges to the exact value slower than the uniformly reduced integration. The gap can be reduced

Table 6.2 Mid-point displacements of cross-ply plate by LRT56 theory using SH85URI elements

Load (+q lb/in^2)	Present (LRT56, $\|w\|/h$)	Reddy [2]	Basar et al. [3]
0.005	0.4303	0.429	0.429
0.01	0.8606	0.858	0.858
0.02	1.7208	1.71	1.717
0.03	2.5803	2.55	2.574
0.04	3.4385	3.37	3.430
0.05	4.2951	4.19	4.285
0.10	8.5413	7.92	8.525
0.25	20.5799	16.17	20.57
0.50	37.0014	24.82	37.03
0.75	48.8280	30.87	48.83
1.0	57.1631	35.69	57.14
2.0	73.7032	49.56	73.64
3.0	80.4214	59.65	80.36
4.0	84.0963	68.00	84.03
5.0	86.4767	75.33	86.41

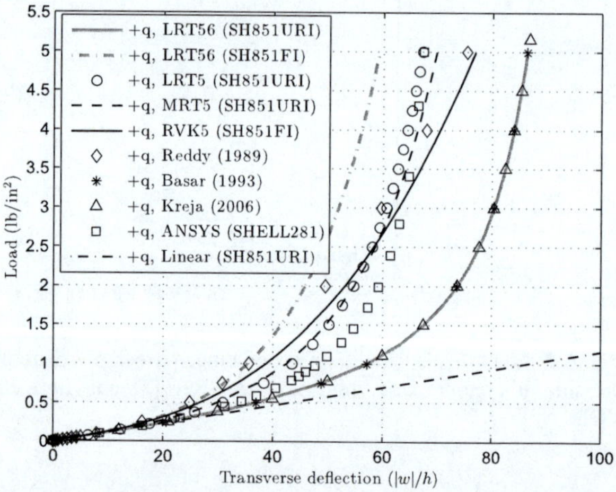

Fig. 6.3 Load-displacement curves of simply supported cross-ply plate under a uniform pressure, reprinted from Ref. [5], copyright 2014, with permission from ELSEVIER

by increasing the number of elements. It can be seen from the figure that results obtained by the model of RVK5 with SH851FI agree very well with those reported by Reddy [2], who implemented the same type of nonlinear theory in the model but based on the TOSD hypothesis. In addition, the results indicate that the commercial software ANSYS predicts similar to the MRT5 and LRT5 models in this example.

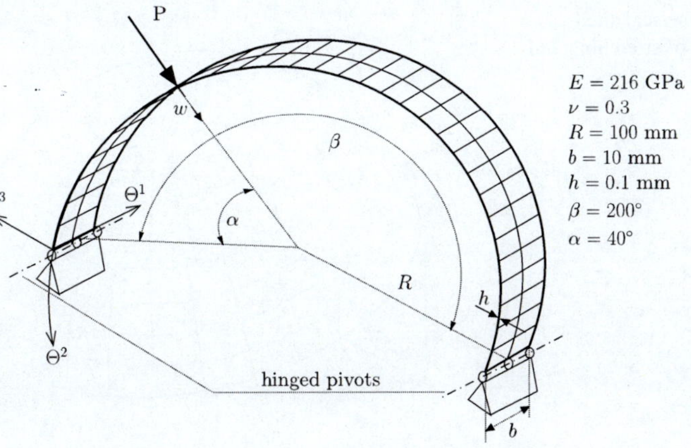

Fig. 6.4 Asymmetrically loaded hinged thin arch, reprinted from Ref. [6], copyright 2014, with permission from ELSEVIER

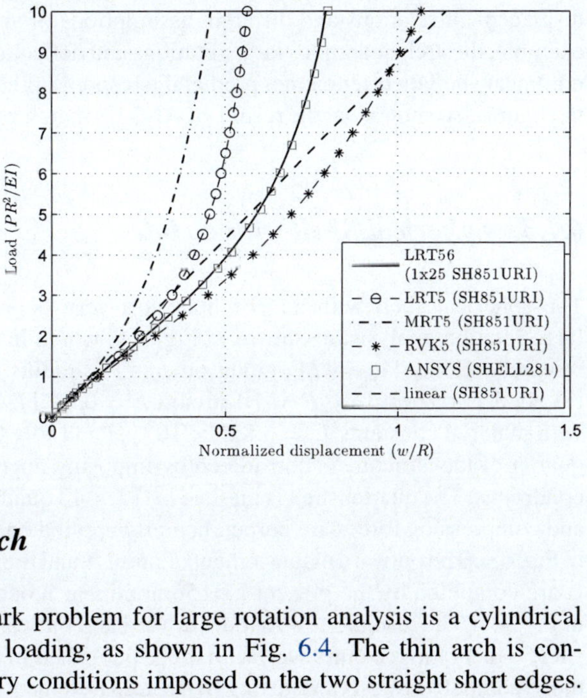

Fig. 6.5 Static response of the asymmetrically loaded hinged thin arch, reprinted from Ref. [6], copyright 2014, with permission from ELSEVIER

6.1.2 Hinged Thin Arch

The second typical benchmark problem for large rotation analysis is a cylindrical thin arch under asymmetric loading, as shown in Fig. 6.4. The thin arch is constrained with hinged boundary conditions imposed on the two straight short edges. The dimensions and material parameters can be found in the figure. The arch is meshed by 25 quadrilateral elements along the hoop direction. An asymmetric point force is applied on the arch located 40° from the left hinged edge. The normalized radial displacements are calculated by various nonlinear models, which are presented and compared in Fig. 6.5.

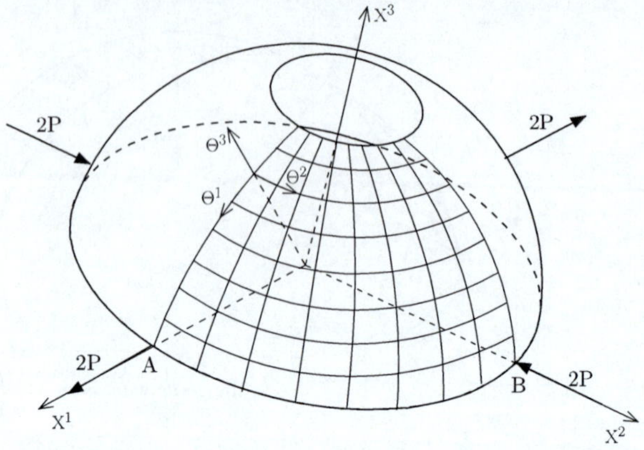

Fig. 6.6 Spherical shell under a pair of stretching and compressing forces

The results indicate that the displacements are considerably different compared to each other. The reasons are that each nonlinear shell theory considers different strain-displacement relations and different assumptions of shell director rotations. In this example, the arch undergoes large rotations and deflections. Therefore, the simplified nonlinear shell theories cannot predict the response precisely, except LRT56. Finally the figure also shows that the results of ANSYS match very well with those of LRT56.

6.1.3 Spherical Shell with a Hole

The spherical shell with an 18° hole is a very popular benchmark problem for large rotation analysis, as shown in Fig. 6.6, which has been investigated by many researchers, like [4, 7–13] among others. The radius and thickness of the spherical shell is respectively $R = 10$ in and $h = 0.04$ in. The material is an isotropic with material constants $E = 6.825 \times 10^7$ psi and $\nu = 0.3$. In the simulation, only a quarter of the structure is considered by imposing appropriate symmetric boundary conditions. The quarter shell is meshed by 12×12 quadrilateral elements. Stretching and compressing forces are perpendicularly applied on the spherical shell, as shown in Fig. 6.6. The outward displacement of point A and the inward displacement at point B are computed by the present LRT56 nonlinear model, with the results presented in Fig. 6.7. The results show that the inward and outward displacements obtained by LRT56 theory agree quite well with those published in the literature, as well as with those computed by ANSYS using SHELL281 elements.

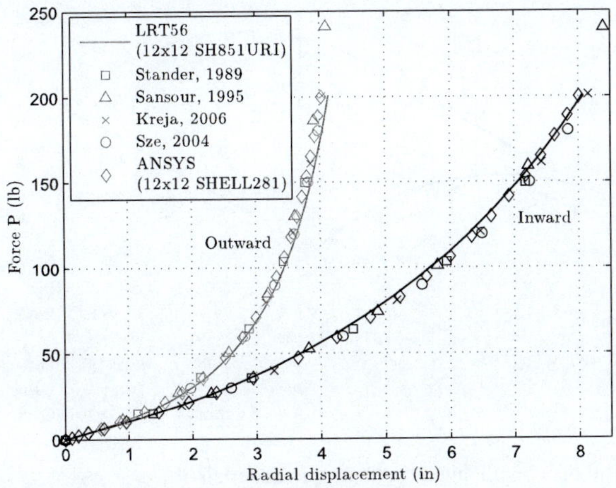

Fig. 6.7 Outward and inward displacements of the spherical shell

6.2 Buckling and Post-buckling Analysis

6.2.1 Hinged Panel with Cross-Ply Laminates

Cylindrical shells subjected to transverse load may appear buckling phenomenon. In such a case, the complex load-displacement curve is solved by arc-length control method. First, cross-ply laminated cylindrical shells are considered, which was earlier studied by Saigal et al. [14], Laschet and Jeusette [15], Brank et al. [16], as shown in Fig. 6.8. The dimensions and material properties can be found in Fig. 6.8. The panel is hinged at the two straight edges, which implies that the rotations about the Θ^1-axis are free. A concentrated tip force is applied on the mid-point of the panel.

In the first configuration, the panel is comprised of three substrate layers, stacked as [90°/0°/90°] and [0°/90°/0°]. The total thickness of the panel is 12.6 mm, and that for each substrate layer is 4.3 mm. Because of the symmetric properties of geometry and stacking sequence, only a quarter of the panel is computed using the symmetric boundary conditions. The panel is meshed by 4×4 SH85URI elements. The load-displacement of the mid-point is illustrated in Fig. 6.9 for the case of [0°/90°/0°] and in Fig. 6.10 for [90°/0°/90°].

The figures indicate that the results of RVK5 and MRT5 are matching very well with those reported in Laschet [15] and Brank [16]. The load-displacement curves of LRT56 and LRT5 are almost the same, which imply that the panel does not occur large rotations. In addition, the results of LRT56 and LRT5 show a stiffer response in the post-buckling stage than those of RVK5 and MRT5. The figures also show that the structure of [90°/0°/90°] has higher stiffness than the case of [0°/90°/0°].

In the second configuration of the cross-ply laminated panel, the total thickness is reduce to 6.3 mm but with the same lay-ups. Due to the reduction of thickness,

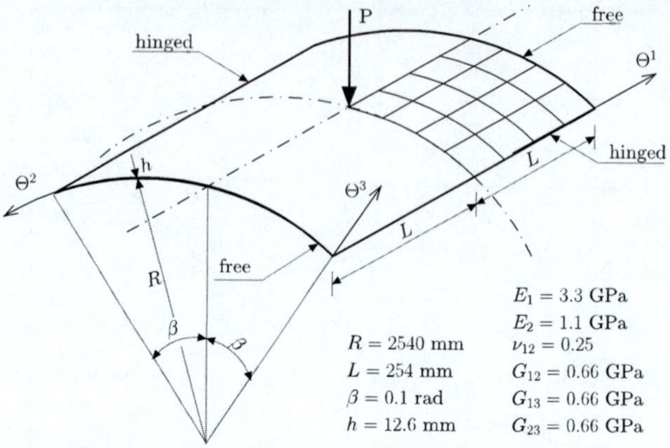

Fig. 6.8 Cylindrical panel with layered orthotropic materials

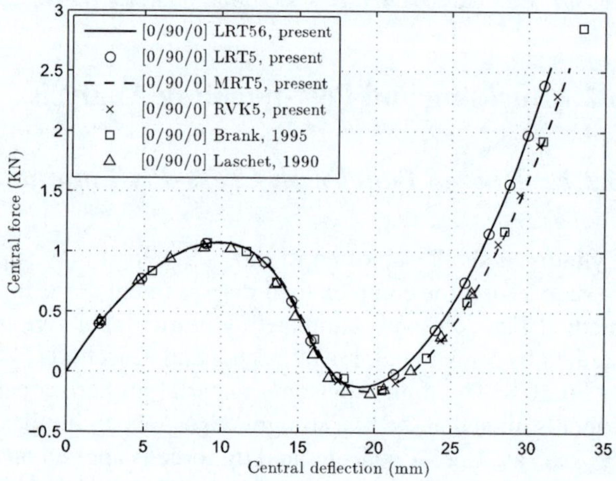

Fig. 6.9 Static response of cross-ply laminated panel with thickness of 12.6 mm and stacking sequence [0°/90°/0°]

a significantly softer load-displacement curves are obtained, as shown in Figs. 6.11 and 6.12, respectively. The panel of [0°/90°/0°] performs a significant, complicated snap-back behavior, while the case of [90°/0°/90°] preforms a relative stable structural response.

6.2.2 *Hinged Panel with Angle-Ply Laminates*

In this simulation, similar structures with same geometries and material properties are investigated. The hinged panel consists of two substrate layers in the configuration of [45°/ − 45°] and [−45°/45°]. Since the stacking sequences are unsymmetrical, the symmetrical boundary conditions in this case will fail in the simulations. However,

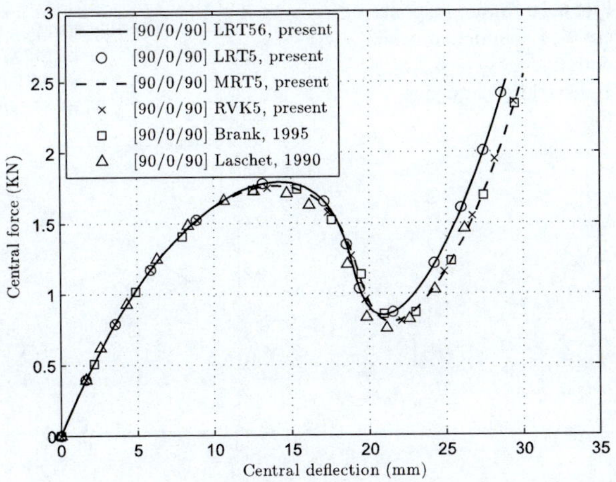

Fig. 6.10 Static response of cross-ply laminated panel with thickness of 12.6 mm and stacking sequence [90°/0°/90°]

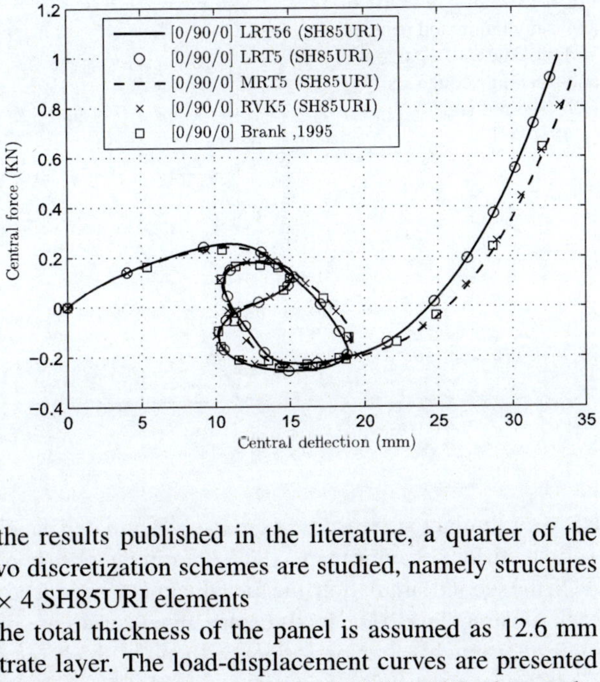

Fig. 6.11 Static response of cross-ply laminated panel with thickness of 6.3 mm and stacking sequence [0°/90°/0°]

in order to compare with the results published in the literature, a quarter of the panel is still considered. Two discretization schemes are studied, namely structures discretized by 1×1 and 4×4 SH85URI elements

In the first simulation, the total thickness of the panel is assumed as 12.6 mm with 6.3 mm for each substrate layer. The load-displacement curves are presented in Fig. 6.13. For the panel with the stacking sequence [45°/ − 45°], the results obtained by MRT5 theory with the mesh of 1×1 eight-node quadrilateral element agrees quite well with those reported in [14]. Here 1×1 eight-node quadrilateral element is equivalent to 2×2 four-node quadrilateral element. The results of LRT56 and MRT5 are matching with each other when the same number of elements are employed. This implies that the structure is undergoing small or moderate rotations.

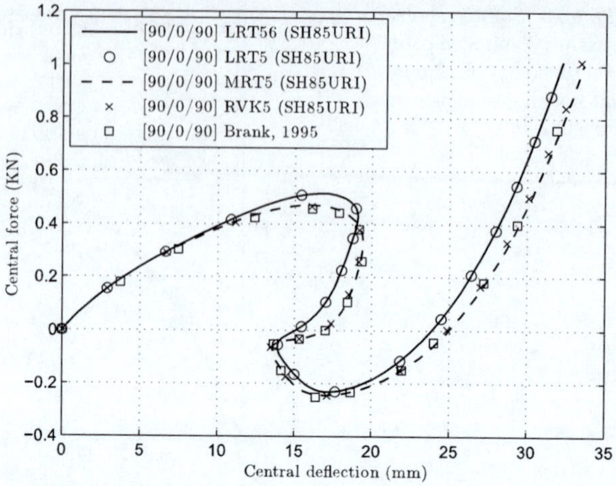

Fig. 6.12 Static response of cross-ply laminated panel with thickness of 6.3 mm and stacking sequence [90°/0°/90°]

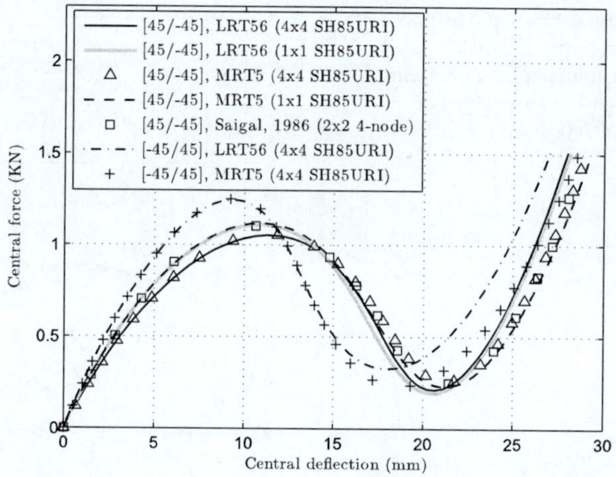

Fig. 6.13 Static response of angle-ply laminated panel with thickness of 12.6 mm and stacking sequences [45°/ − 45°] and [−45°/45°]

It is very similar to the results of cross-ply laminated panel that LRT56 theory predicts a slight stiffer behavior than MRT5 does in the post-buckling range.

In the second simulation, the total thickness is reduced to 6.3 mm with 3.15 mm for each substrate layer. The load-displacement curves are presented in Fig. 6.14, which indicate a much softer performance than the thick ones. The structures exhibits a complex snap-through and snap-back load-deflection path. It is obviously seen that the structure of [−45°/45°] has stiffer response than that of [45°/ − 45°].

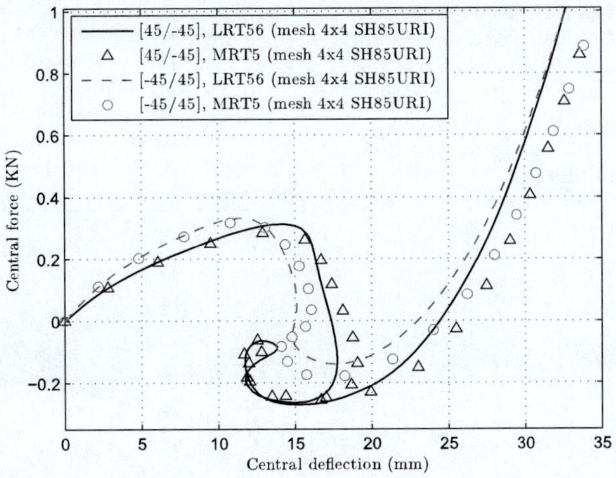

Fig. 6.14 Static response of angle-ply laminated panel with thickness of 6.3 mm and stacking sequences [45°/ − 45°] and [−45°/45°]

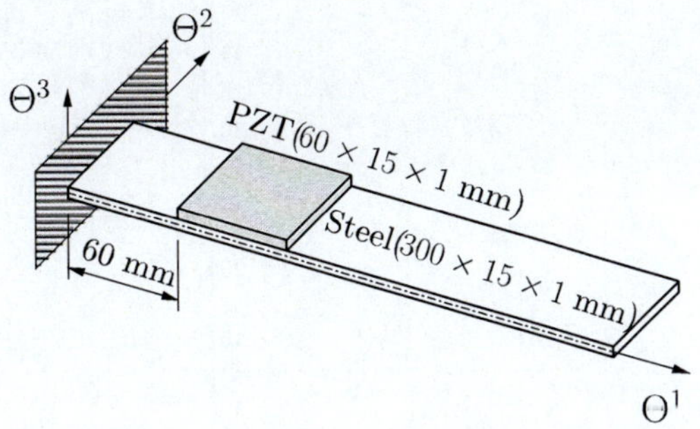

Fig. 6.15 Cantilevered beam with one piezoelectric patch bonded

6.3 Geometrically Nonlinear Analysis of Smart Structures

6.3.1 Cantilevered Smart Beam

From this subsection, thin-walled structures laminated with piezoelectric pathes are studied. First, a cantilevered beam structure integrated with piezoelectric patch is considered, which was proposed by Yi et al. [17], as shown in Fig. 6.15. The dimensions of the host beam structure are $300 \times 15 \times 1$ mm^3, while those of the PZT patch are $60 \times 15 \times 1$ mm^3. The PZT is placed 60 mm away from the cantilevered end. The material properties for the host structure and the PZT are listed in Table 6.3. The same value of piezoelectric coupled parameters d_{31} and d_{32} are used in this report with those in [18]. However, the piezoelectric parameter is a slight difference from those presented in [17]. The electric potential of the piezoelectric patch at the bonded surface assumes to be 0 V, while that of the piezoelectric patch at the outer surface is imposed by a physical equipotential condition.

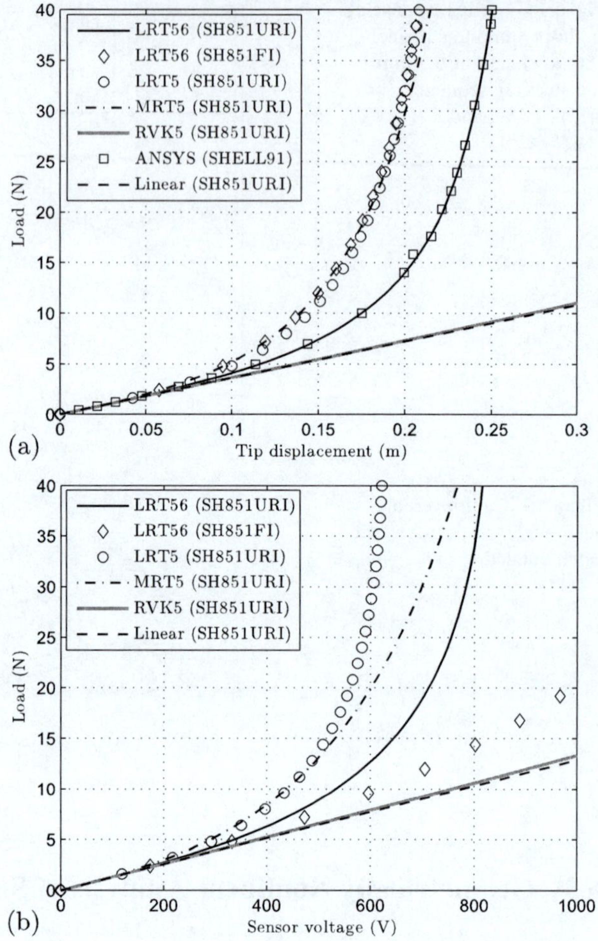

Fig. 6.16 Static response of the cantilevered smart beam: **a** tip displacement, **b** sensor output voltage, reprinted from Ref. [5], copyright 2014, with permission from ELSEVIER

6.3.1.1 Static Analysis

The smart beam is meshed by 10 eight-node quadrilateral elements along the length direction. For the purpose of comparison, both uniformly reduced integration and full integration schemes are used. A tip load up to 40 N is applied on the free end. Using various nonlinear models, the static tip displacements are computed, as well as the voltage generated by piezoelectric patch, shown respectively in Fig. 6.16a, b.

The figure shows that the load-displacement curve of LRT56 model using SH851URI elements has an excellent agreement with that obtained by ANSYS. Regarding to the displacements and sensor voltages, there are large discrepancies between nonlinear models, e.g. LRT56, LRT5, MRT5 and RVK5. If structures undergoing large displacements and rotations, the simplified nonlinear theories (LRT5, MRT5 and RVK5) invalid for the prediction of static behavior. The reasons have

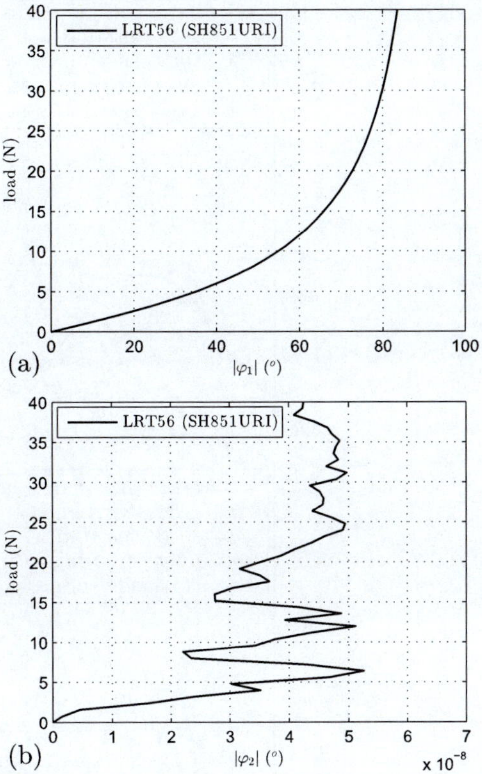

Fig. 6.17 Maximum rotations at centerline nodes of cantilevered smart beam: **a** rotation φ_1, **b** rotation φ_2, reprinted from Ref. [5], copyright 2014, with permission from ELSEVIER

been stated in Sect. 6.1.1. Furthermore, the figure indicates that the load-displacement curves obtained by LRT56 using SH851FI and SH851URI elements have large gaps. This is because the SH851FI elements exhibit membrane and shear locking phenomena, while SH851URI elements avoid these locking effects.

The maximum values of the two rotational DOFs at the centerline nodes are computed and shown in Fig. 6.17a and b, respectively. The figures illustrate that the rotations about the Θ^2-axis are very large. It reaches 80° when the force is around 30 N. However, the rotations about the Θ^1-axis are almost zero, due to the fact that the beam is bent only in one direction without any torsional effect. The small deviation of φ_2 is caused by the error margin of the numerical computation.

6.3.1.2 Dynamic Analysis

In the dynamic analysis, two discretization methods are considered, namely 5 × 1 and 10 × 1 eight-node shell elements. A step load with the amplitude of 10 N is applied on the tip point. For linear dynamic calculation, the Newmark method is used with a time step of 1×10^{-3} s. For geometrically nonlinear dynamic response,

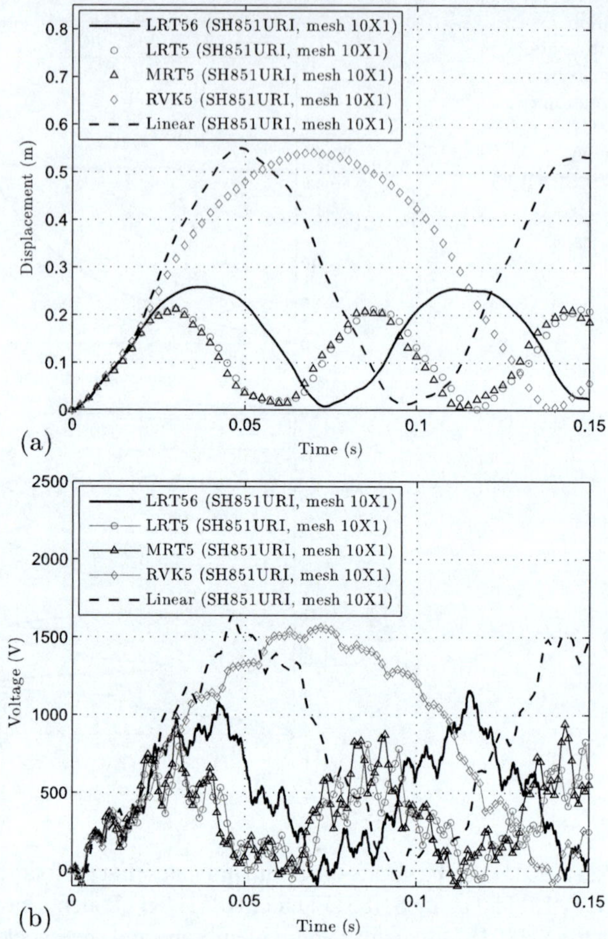

Fig. 6.18 Dynamic response of cantilevered beam using various shell theories: **a** tip displacement, **b** sensor output voltage, reprinted from Ref. [19], copyright 2013, with permission from IOP

the Newmark method with a time step of 5×10^{-6} s is considered to solve the nonlinear models with SH851URI elements, the CDA method with a time step of 1×10^{-7} s is employed to solve the models with SH851FI elements. The transient response of tip displacement and sensor voltage is obtained by various nonlinear models using SH851URI elements, shown in Fig. 6.18a and b, respectively.

From Fig. 6.18a, it implies that LRT56 predicts response stiffer than RVK5, but softer than MRT5 and LRT5. The time histories of various nonlinear models are matching the static displacements in Fig. 6.16. Because of large rotations occur in the smart beam, these simplified nonlinear models fail to predict the dynamic response precisely. The analysis on rotations of shell director under a quasi-statically applied tip force of 10 N is presented in Fig. 6.17a, which shows a maximum rotation over

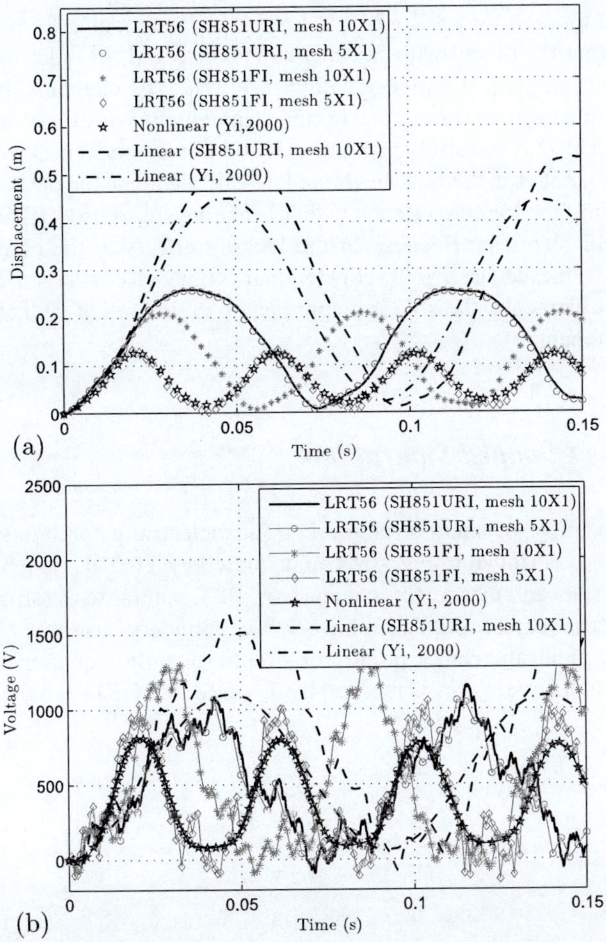

Fig. 6.19 Dynamic response of cantilevered beam using various meshes and integration schemes: **a** tip displacement, **b** sensor output voltage, reprinted from Ref. [19], copyright 2013, with permission from IOP

50° existing in the structure. The results in Fig. 6.18b show that the linear theory overpredicts the sensor output voltage, since the linear model does not take into account the stress stiffening effects. In addition, the figure of sensor output voltage indicates that the amplitude of time history obtained by LRT56 model is larger than those obtained by MRT5 and LRT5. This can be explained that the larger deformation the higher output voltage.

The next comparisons are among the results of LRT56 model meshed by 5×1 and 10×1 elements in consideration of both uniformly reduced integration and full integration, as presented in Fig. 6.19a for the displacements and (b) for the sensor output voltages. The time histories in Fig. 6.19a show that the transient response of

LRT56 model meshed by 5 × 1 or 10 × 1 SH851URI elements is almost identical. This implies that the model with 5 elements is converged. In order to compare the results of Yi et al. [17], the model with 5 × 1 SH851FI elements are considered in the simulation, whose results are in good agreement with those in [17]. In the reference, Yi et al. [17] used fully geometrically nonlinear model meshed by 5 twenty-node solid elements for the host beam and 1 solid element for piezoelectric patch without avoiding the locking effects, which is equivalent to use LRT56 model with 5 × 1 SH851FI elements. Because of the locking effects in the computation, the nonlinear model meshed by 5 SH851FI elements is not well converged. By increasing the number of elements, the solution approaches to the one of LRT56 model with SH851URI elements.

6.3.2 *Fully Clamped Smart Plate*

The second example of smart structure is a piezoelectric patch bonded plate with four edges fixed, as shown in Fig. 6.20, first studied by Yi et al. [17]. The host plate is made up of steel and piezoelectric is made of PZT, similar to the smart beam. The material properties can be found in Table 6.3. The boundary conditions and geometry are symmetric about the center point. For simplicity, only a quarter of the plate is considered, and the structures is meshed by 5 × 5 SH851URI shell elements.

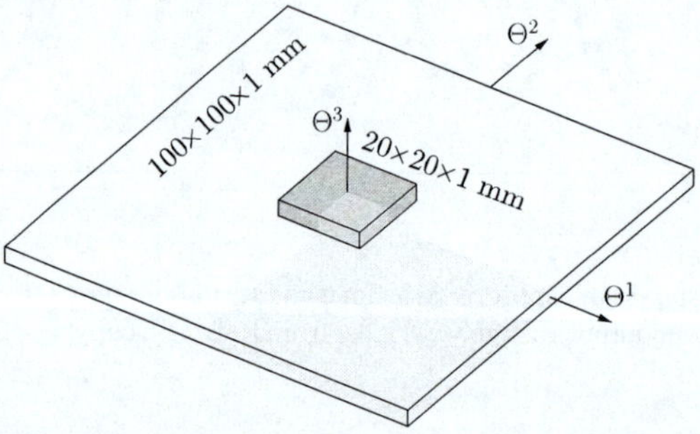

Fig. 6.20 Fully clamped plate with one piezoelectric patch centrally bonded

Table 6.3 Material properties of the cantilevered smart beam

Steel	PZT
$E = 197$ GPa	$E = 67$ GPa
$\nu = 0.33$	$\nu = 0.33$
$\rho = 7900$ kg/m^3	$\rho = 7800$ kg/m^3
	$d_{31} = d_{32} = -1.7119 \times 10^{-10}$ C/N
	$\epsilon_{33} = 2.03 \times 10^{-8}$ F/m

6.3.2.1 Static Analysis

For the static analysis, the smart plate is subjected to a uniformly distributed surface load with the maximum value 2×10^7 Pa. In the simulation, the electrodes of piezoelectric patches are short circuited, meaning that the sensor electric voltage will not influence the deformation of the plate. The mid-point displacements are computed by different nonlinear theories, which are presented in Fig. 6.21a. The figure shows that the load-deflection curves obtained by various models as well as the commercial software ANSYS are almost identical. This is because the fixed boundary conditions, the smart plate cannot undergo large rotations, but undergo only moderate rotations. It can be demonstrated by observation of the rotations in the plate, φ_1 and φ_2, which is calculated by LRT56 theory using SH851URI elements in a loading condition of 2×10^7 Pa, as shown in Fig. 6.22a and b, respectively. The figure shows that the rotations jump suddenly from zero to a maximum value, then decrease to zero at the mid-point. The sensor output voltages under different load levels are shown in Fig. 6.21b. The results show that the linear theory overpredicts the sensor output voltage, because it does not account for the stress stiffening effects. Since the plate undergoes only moderate rotations, the results obtained by all nonlinear theories are in very good agreement.

6.3.2.2 Dynamic Analysis

The dynamic response is investigated under a uniformly distributed step pressure with the amplitude of 2×10^4 N/m^2. Both the linear and nonlinear displacement time histories are solved by Newmark method. A time step of 1×10^{-5} s is used for the linear case, and 1×10^{-7} s for the nonlinear case. The dynamic response of the mid-point displacement and sensor output voltage both for linear and nonlinear simulations is presented in Fig. 6.23a and b, respectively. The results indicate that the linear and nonlinear transient response is very similar to each other. This is because the plate is undergoing small displacements and rotations, meaning that only weak nonlinear effects are imposed. The present time histories of both displacement and sensor output voltage agree excellently with those presented by Lentzen and Schmidt [18]. The present results are obtained by LRT56 and RVK5 models in the FOSD hypothesis, while those of Lentzen and Schmidt [18] are achieved by FOSD MRT5 theory. This is because weak nonlinear effects and only moderate rotations are occurring in the plate. Due to the fact that the linear theory does not take into account for the von Kármán stress stiffening effect, the time histories of linear theory have slightly larger amplitudes.

To present nonlinear phenomenon, the magnitude of the surface load increases to 2×10^5 N/m^2. The dynamic response curves are calculated by linear and nonlinear theories are presented in Fig. 6.24. Now the amplitudes of the displacements are in the order of the magnitude of the plate thickness. Therefore, a big difference can be observed between the linear and nonlinear vibrations. This is due to the stress stiffening effect. Since the linear theory does not account for this effect, it overpredicts

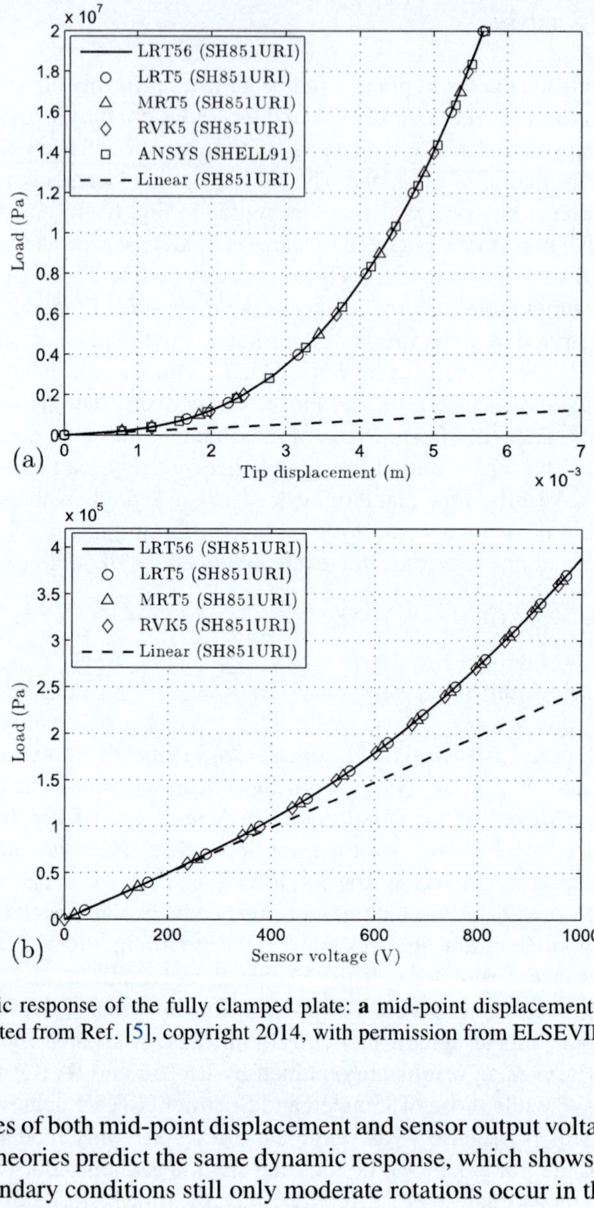

Fig. 6.21 Static response of the fully clamped plate: **a** mid-point displacement, **b** sensor output voltage, reprinted from Ref. [5], copyright 2014, with permission from ELSEVIER

the amplitudes of both mid-point displacement and sensor output voltage. The RVK5 and LRT56 theories predict the same dynamic response, which shows that due to the clamped boundary conditions still only moderate rotations occur in the plate.

6.3.3 Fully Clamped Cylindrical Smart Shell

The first smart shell structure considered in this report is a fully clamped cylindrical shell bonded with piezoelectric patch on the center, as shown in Fig. 6.25. The

(a)

(b)

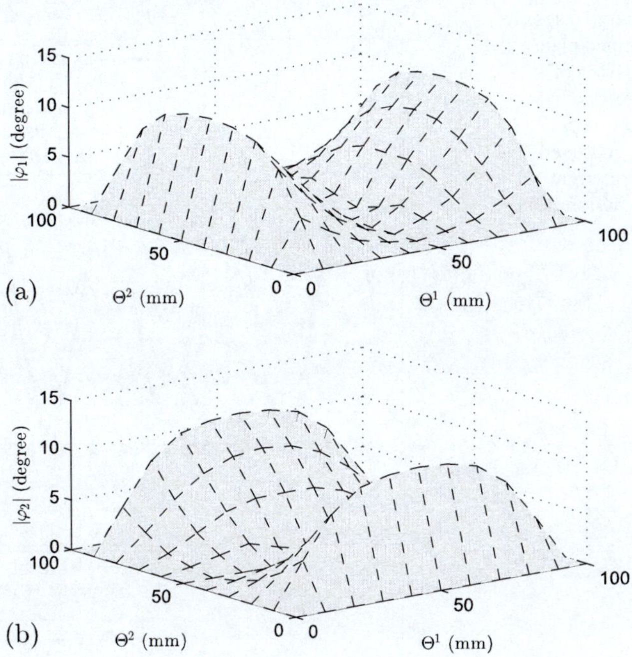

Fig. 6.22 Rotations of the plate under a pressure of 2×10^7 Pa: **a** rotation φ_1, **b** rotation φ_2, reprinted from Ref. [5], copyright 2014, with permission from ELSEVIER

structure was first proposed and studied by Yi et al. [17]. The smart cylindrical shell consists of a host structure made up of orthotropic material and a piezoelectric patch as actuator or sensor. The material properties are given in Table 6.4. The direction of the fiber reinforcement of the host shell structure is along the Θ^1-axis. A symmetric boundary conditions and geometry are imposed. Thus, a quarter of the cylindrical shell is taken into account in the simulations, discretized by 8×4 SH851URI elements along the Θ^1- and Θ^2-axis.

6.3.3.1 Static Analysis

A uniformly distributed surface load pointing outward is applied on the shell structure, with the maximum value of 2×10^7 Pa. In the analysis, the sensor electrodes are short circuited. The mid-point displacements of geometrically nonlinear models are calculated and presented in Fig. 6.26a. The results indicate that the load-displacement curves of RVK5, MRT5, LRT5, LRT56 and ANSYS are almost identical. This can be explained by that the shell structure undergoes only moderate rotations due to the fixed boundary conditions. The rotations, φ_1 and φ_2, can be observed in Fig. 6.27a and b, respectively. The results indicate slightly that the load-displacement curves of LRT56/LRT5 are different from those of MRT5/RVK5. This difference is only

Fig. 6.23 Dynamic response of the fully clamped plate under a step pressure of 2×10^4 Pa: **a** mid-point displacement, **b** sensor output voltage, reprinted from Ref. [6], copyright 2014, with permission from ELSEVIER

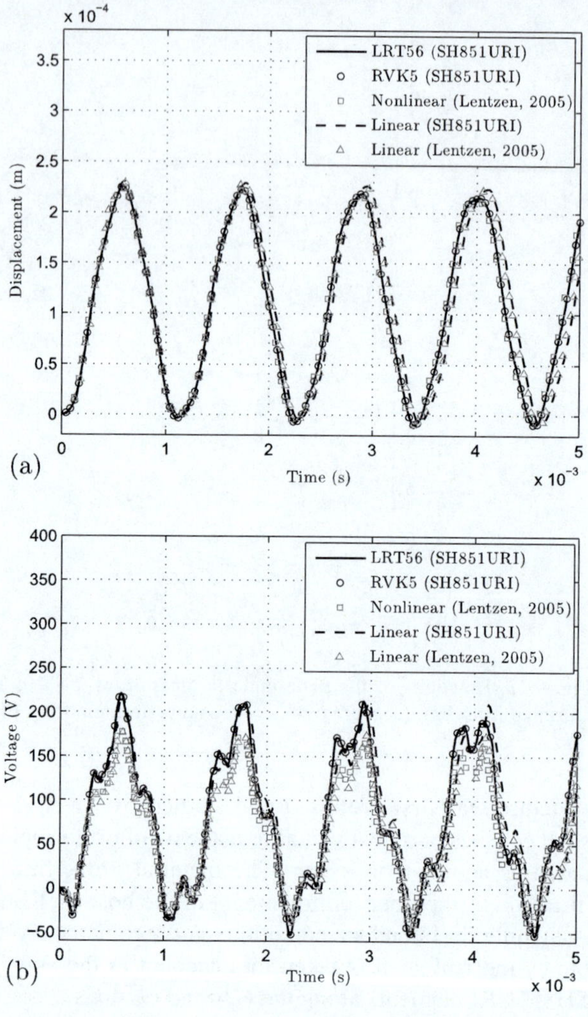

from nonlinear strain-displacement terms. From the sensor output voltage results, given in Fig. 6.26b, it shows identical load-voltage curves of all nonlinear models are achieved.

6.3.3.2 Dynamic Analysis

In the dynamic analysis, the Newmark method is employed for solving linear and nonlinear dynamic equations. For the linear case, the time step is chosen 1×10^{-5} s, while for the nonlinear case, the time step is 1×10^{-7} s. A uniformly step surface force is applied on the shell structure. The transient response of the mid-point dis-

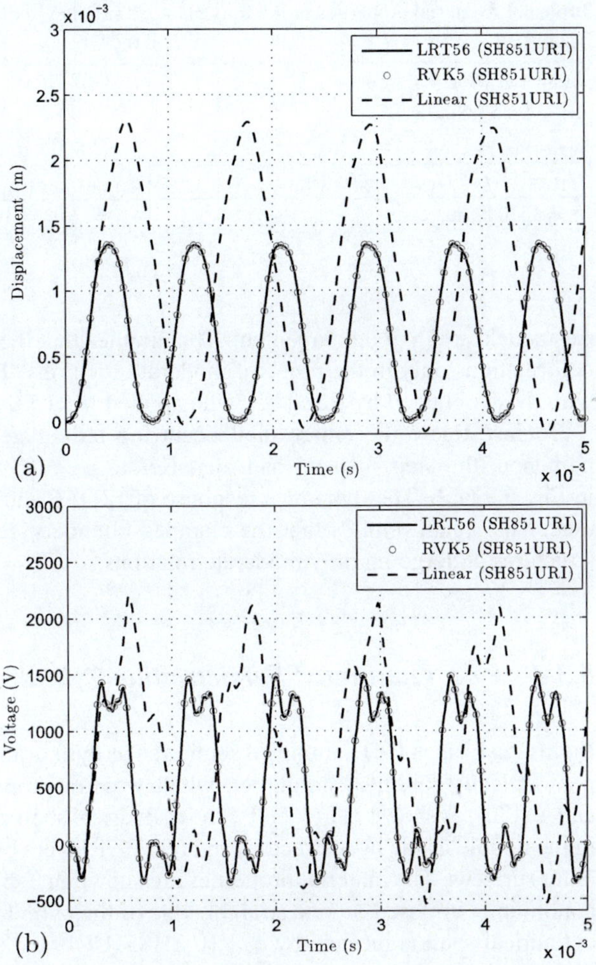

Fig. 6.24 Dynamic response of the fully clamped plate under a step pressure of 2×10^5 Pa: **a** mid-point displacement, **b** sensor output voltage, reprinted from Ref. [6], copyright 2014, with permission from ELSEVIER

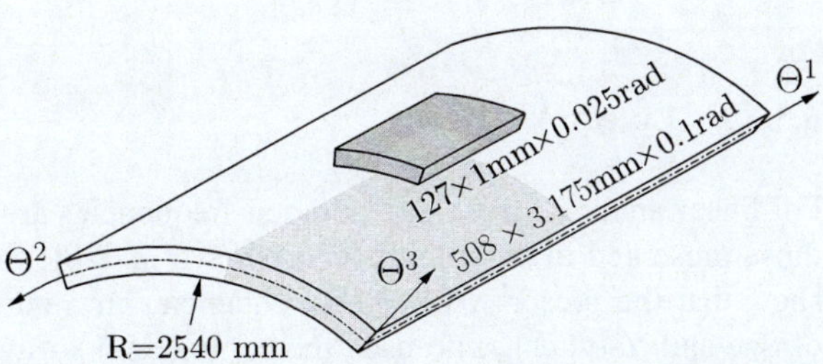

Fig. 6.25 Fully clamped cylindrical shell with one piezoelectric patch centrally bonded

placement and the sensor output voltage is computed and presented in Fig. 6.28a and b, respectively. Two different nonlinear models are studied, LRT56 and RVK5. The time histories of the displacement and sensor voltage obtained by these two nonlin-

Table 6.4 Material properties of the fully clamped smart cylindrical shell

Host shell	PZT
$E_1 = 124$ GPa	$E = 67$ GPa
$E_2 = 96.53$ GPa	$\nu = 0.33$
$\nu_{12} = \nu_{23} = 0.34$	$\rho = 7800$ kg/m^3
$G_{12} = G_{13} = G_{23} = 6.205$ GPa	$d_{31} = d_{32} = -1.7119 \times 10^{-10}$ C/N
$\rho = 1520$ kg/m^3	$\epsilon_{33} = 2.03 \times 10^{-8}$ F/m

ear models are in good agreement. This implies that the cylindrical shell undergoes deformations only in the range of moderate rotations. The results agree excellently with those obtained by Ref. [18], who applied MRT5 theory in the computation.

To investigate the smart shell occurring strong geometrically nonlinear phenomenon, the step surface load increases to 6×10^5 Pa, with the results shown in Fig. 6.29a, b. The dynamics response of LRT56 and RVK5 models is still identical. This again implies that the clamped boundary conditions mostly permit the structures undergoing only moderate rotations.

6.3.4 *PZT Laminated Semicircular Cylindrical Shell*

In this analysis, a PZT laminated semicircular cylindrical shell is considered for the geometrically nonlinear simulation, which was first proposed and calculated by Tzou and Ye [20], as shown in Fig. 6.30. The cylindrical shell consists of one metallic layer in the middle as the host structure and two PZT layers bonded on the both inner and outer surfaces. The material properties are shown in Table 6.5. A clamped boundary condition is imposed on one straight edge of the cylindrical shell. The semicircular cylindrical shell is meshed by 1×10 SH851URI elements along the Θ^1- and Θ^2-direction, respectively.

6.3.4.1 Linear Analysis

For linear analysis, the first five eigen-frequencies are analyzed and compared with those presented in the reference, as shown in Table 6.6. From the table, the results show that the present five eigen-frequencies in a good agreement with the results of Sze and Yao [21], who used the commercial software ABAQUS and self coded program. It can be seen that a large discrepancy between the present results and those of Ref. [20]. The reason might be that a different Young's modulus of the metal are employed in the reference.

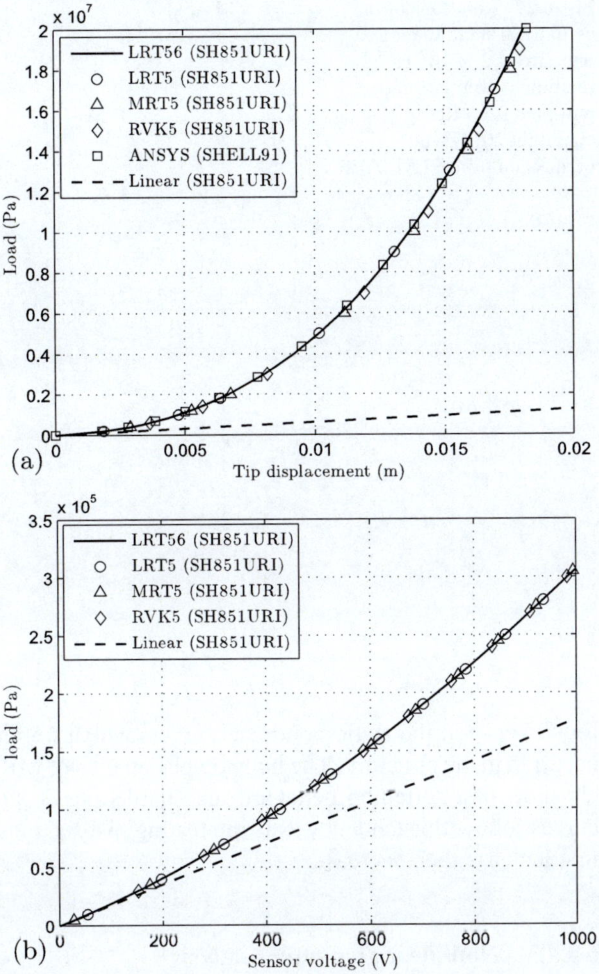

Fig. 6.26 Static response of the fully clamped smart cylindrical shell: **a** mid-point displacement, **b** sensor output voltage, reprinted from Ref. [5], copyright 2014, with permission from ELSEVIER

6.3.4.2 Nonlinear Static Analysis

In this simulation, the nonlinear static analysis are studied for both tip point displacements and sensor output voltages of inner layer. Applying a concentrated force along the positive hoop direction, the load-displacement and load-voltage curves are obtained and presented in Fig. 6.31. It can be seen that the RVK5 nonlinear model performs softer behavior than the linear model. The figures show that the load-displacement and load-voltage curves in the hoop direction obtained by MRT5 and LRT5 models are quit similar to each other. This confirms the conclusion in Sect. 3.5 that the LRT5 theory is restricted to the range of moderate rotations even though fully geometrically nonlinear relations are used. By comparison of the results obtained by RVK5, MRT5, LRT5 and LRT56 in Fig. 6.31a, large deviations can be

Fig. 6.27 Rotations of the cylindrical shell under a pressure of 2×10^7 Pa: **a** rotation φ_1, **b** rotation φ_2, reprinted from Ref. [5], copyright 2014, with permission from ELSEVIER

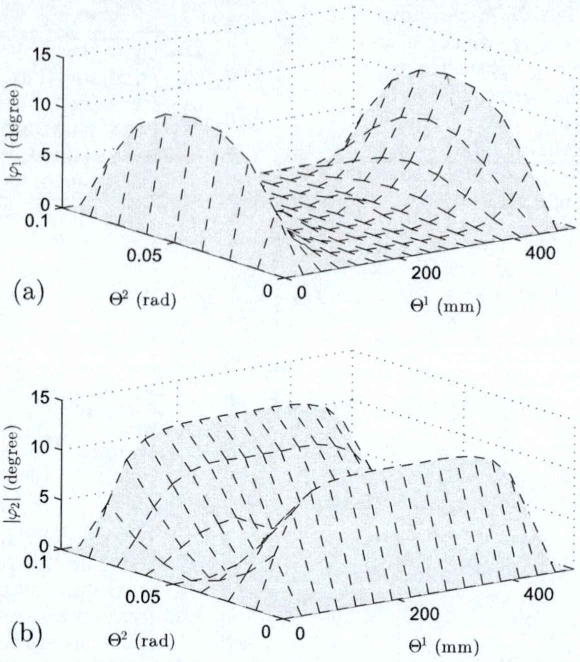

found between the static nonlinear curves, which implies that large rotations have occurred in the structure. The hoop displacements of MRT5, LRT5 and LRT56 models show first softening behavior, then hardening at large loads. The load-voltage curves follow this tendency. It is interesting that the radial displacements of LRT56 first increase then decrease.

6.3.4.3 Nonlinear Dynamic Analysis

In this part, the nonlinear dynamic analysis of the semicircular cylindrical shell is investigated. A step load with the amplitude of $F_1 = 50$ N is applied on the tip point at the free end along the positive hoop direction. The dynamic response of the tip displacement in the hoop and radial directions, as well as the output voltage of the inner PZT layer is calculated using Newmark method with a time step 1×10^{-3} s for the linear case and 1×10^{-4} s for the nonlinear case, which is respectively shown in Fig. 6.32a–c.

In the nonlinear static analysis, it has been shown that MRT5 and LRT5 models perform very similar. Therefore, the time histories obtained by MRT5 and LRT5 models have very small differences. Because of the softening behavior of LRT56, LRT5 and MRT5 models, larger amplitudes of dynamic vibrations in hoop direction are achieved than those of linear model. This tendency of dynamic response is in accordance with that of static response as shown in Fig. 6.31a. In addition, from

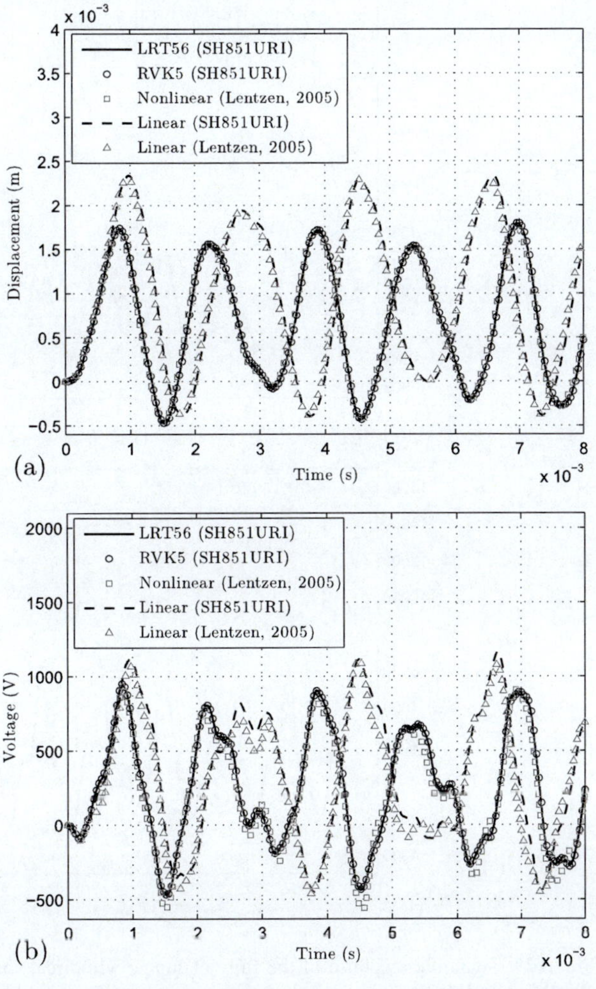

Fig. 6.28 Dynamic response of the fully clamped cylindrical shell under a step pressure of 6×10^4 Pa: **a** mid-point displacement, **b** sensor output voltage, reprinted from Ref. [19], copyright 2013, with permission from IOP

Figs. 6.31 and 6.32, it shows that in a certain range of load, up to around 160 N for the hoop direction and up to around 75 N for the radial direction, MRT5 overpredicts the static tip deflections and sensor output voltages. These observations are also confirmed by a comparison of the frequencies of the transient response of the tip deflection and the sensor output voltage in Fig. 6.32, which shows that the stiffest response is predicted by RVK5, followed, in this sequence, by the linear theory, LRT56 and MRT5. Additionally, the vibrations in radial direction obtained by MRT5, LRT5 and LRT56 models show a superimposed feature of multi-signals. They lead to smaller amplitudes than those of linear theory, which is in contrast to the tendency in the static analysis. A analogous phenomenon occurs in the graph of the inner layer sensor output voltage predicted by MRT5 and LRT5.

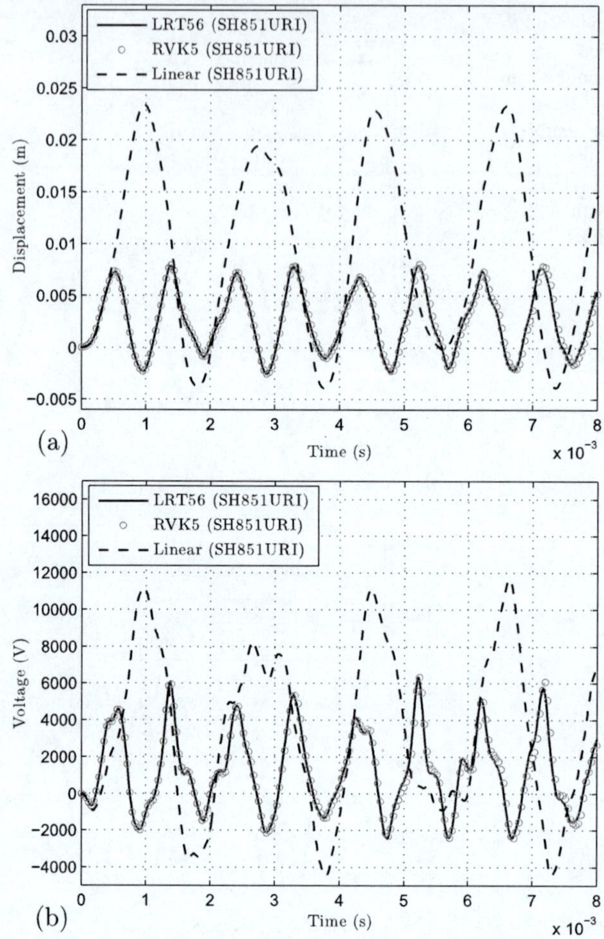

Fig. 6.29 Dynamic response of the fully clamped cylindrical shell under a step pressure of 6×10^5 Pa: **a** mid-point displacement, **b** sensor output voltage, reprinted from Ref. [19], copyright 2013, with permission from IOP

By comparison of all the nonlinear dynamic vibrations, it observes that large discrepancies exist between those obtained various nonlinear models. This means that the semicircular shell is undergoing large displacements and rotations. Therefore, LRT56 model predicts the most accurate dynamic response, rather than the simplified nonlinear shell theories RVK5, MRT5 and LRT5.

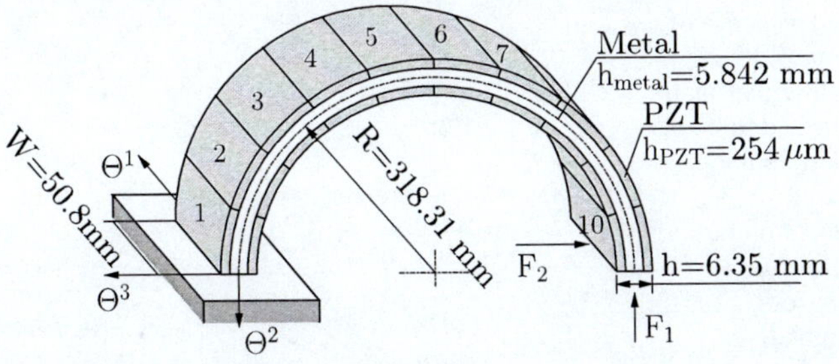

Fig. 6.30 PZT laminated semicircular cylindrical shell

Table 6.5 Material properties of the PZT laminated semicircular cylindrical shell

Host shell	PZT
$E = 68.95$ GPa	$E = 63$ GPa
$\nu = 0.3$	$\nu = 0.3$
$\rho = 7750$ kg/m^3	$\rho = 7600$ kg/m^3
	$d_{31} = d_{32} = -1.79 \times 10^{-10}$ C/N
	$\epsilon_{33} = 1.65 \times 10^{-8}$ F/m

Table 6.6 First five eigen-frequencies of the PZT laminated semicircular cylindrical shell (Hz)

	Mesh	Mode 1	Mode 2	Mode 3	Mode 4	Mode 5
Present	1 × 10 SH851URI	3.7199	5.8530	11.7782	33.4356	40.3605
Sze and Yao [21]	1 × 10 S9R5	3.7475	5.8971	11.856	33.634	40.626
Sze and Yao [21]	2 × 20 4-node	3.6810	5.8041	11.691	33.231	40.450
Tzou and Ye [20]	2 × 10 Triangular	8.17	25.66	86.93	194.14	346.08

6.4 Electroelastic Nonlinear Analysis of Smart Structures

6.4.1 Validation Test

In most of cases, the piezoelectric smart structures are under strong electric field in order to achieve large actuation forces. In such a case, electroelastic materially nonlinear effect should be considered during the analysis of piezoelectric integrated smart structures. To verify the current nonlinear finite element model under strong driving electric field, a cantilevered bimorph beam is analyzed, which was first studied by Wang et al. [22]. The bimorph beam consists of two soft PZT (3203HD) layers, with the dimensions of 35 × 7 × 0.5 mm^3 for each layer, as shown in Fig. 6.33.

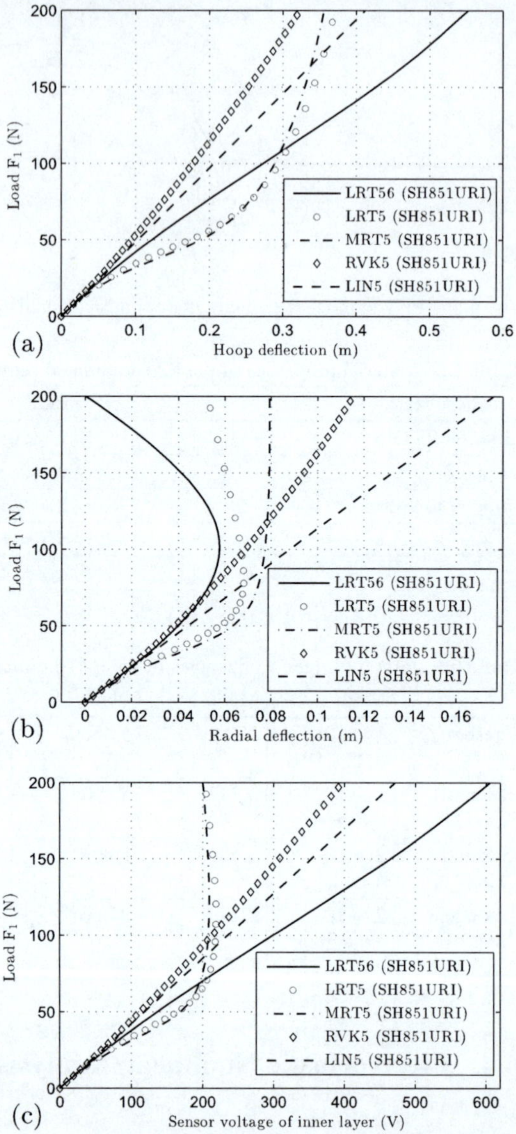

Fig. 6.31 Static response of the PZT laminated semicircular cylindrical shell under a concentrated force in the hoop direction: **a** hoop deflection, **b** radial deflection, **c** sensor output voltage of the inner PZT layer, reprinted from Ref. [6], copyright 2014, with permission from ELSEVIER

The piezoelectric layers in the bimorph beam have opposite polarizations, pointing outwards and perpendicular to the mid-surface. The two substrate layers assume perfectly bonded at both the negative electrodes that are connect to ground. The material parameters for 3203HD are given in Table 6.7. The bimorph beam is meshed by 2×10 eight-node quadrilateral elements with 5 mechanical DOFs at each node and 1 electric DOF at each element. An equal electric potential is applied on both top and bottom surfaces of 3203HD patches. The displacement of tip point is com-

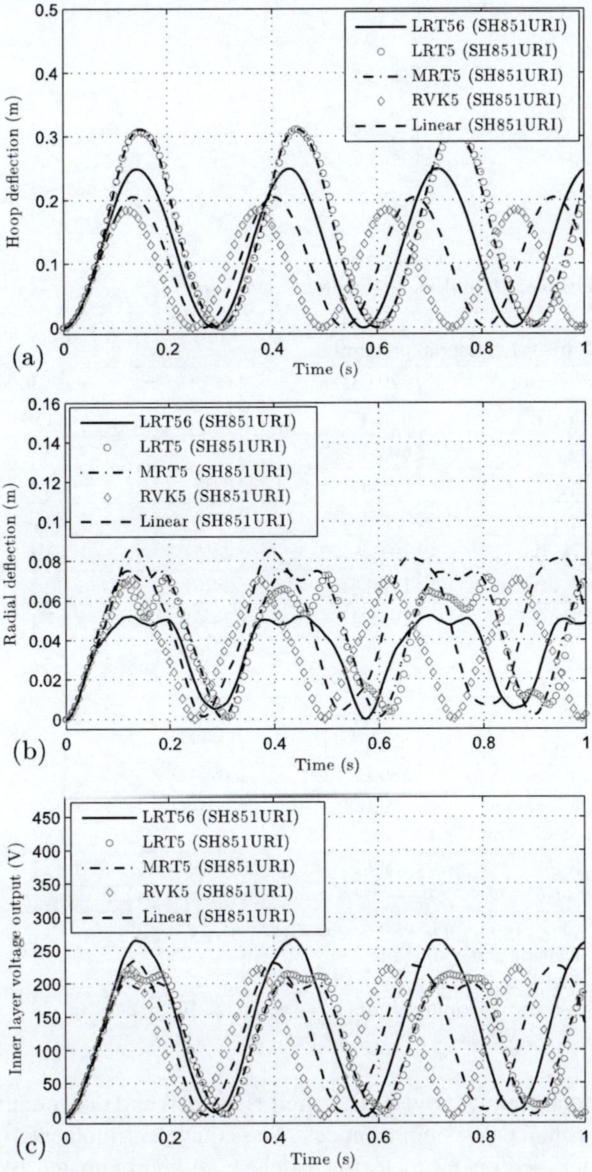

Fig. 6.32 Dynamic response of the PZT laminated semicircular cylindrical shell under a step tip force of 50 N: **a** hoop deflection, **b** radial deflection, **c** sensor output voltage of the inner PZT layer, reprinted from Ref. [6], copyright 2014, with permission from ELSEVIER

puted by models of LIN5WE and LIN5SE, with the load-displacement curves shown in Fig. 6.34.

Only including the electroelastic materially nonlinear phenomenon, The results of LIN5WE and LIN5SE models are compared with the experimental and linear theoretical static data in Wang et al. [22], as well as those obtained by Abaqus. The Abaqus results are obtained through the discretization of 2 × 10 S8R5 elements (8-

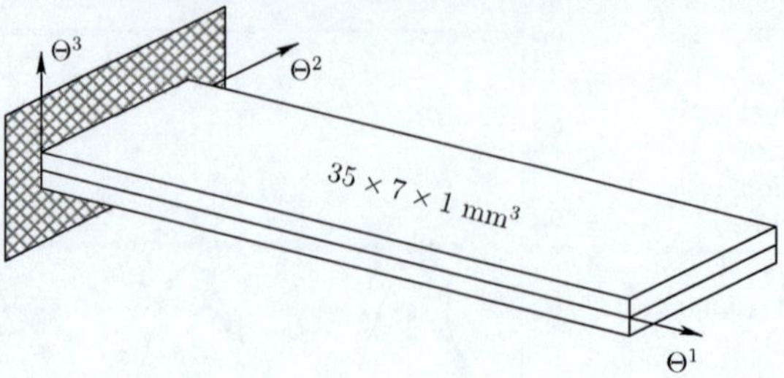

Fig. 6.33 Cantilevered bimorph beam

Table 6.7 Material properties

Properties	3203HD[a,b]	G-1195[b,c]	T300/976	Aluminum
Y_1 (GPa)	60.24	63	150	70
Y_2 (GPa)	60.24	63	9	70
ν_{12}	0.253	0.28	0.3	0.3
ν_{23}	0.494	0.28	0.3	0.3
G_{12} (GPa)	20.04	24.61	7.1	26.92
G_{23} (GPa)	19.084	24.61	2.5	26.92
G_{13} (GPa)	19.084	24.61	7.1	26.92
d_{31} ($\times 10^{-12}$ m/V)	−320	−254	/	/
d_{32} ($\times 10^{-12}$ m/V)	−320	−254	/	/
ϵ_{33}/ϵ_0 [d]	3800	1694.9	/	/
β_{331} ($\times 10^{-18}$ m^{2}V^{-2})	−520	−165	/	/
β_{332} ($\times 10^{-18}$ m^{2}V^{-2})	−520	−165	/	/

[a] CTS Corporation: http://www.ctscorp.com/
[b] Kapuria and Yasin [24]
[c] Yao et al. [25]
[d] Electrical permittivity of air, $\epsilon_0 = 8.85 \times 10^{-12}$ F/m

node doubly curved thick shell elements) and under equivalent moments distributed at the corresponding nodes. The equivalent moments due to driving electric field imposed on piezoelectric patches are approximated by integral of the product of longitudinal stress and moment arm over the cross-section, where the electroelastic materially nonlinear constitutive relations given in (4.73) are considered at different levels of electric fields. The present results of LIN5WE and LIN5SE models agrees excellently with those in the literature and obtained by Abaqus.

The next example for validation study is a simply supported smart plate bonded with two piezoelectric substrate layers, as shown in Fig. 6.35. The host plate is made of aluminum and piezoelectric material is chosen as G-1195. The dimensions

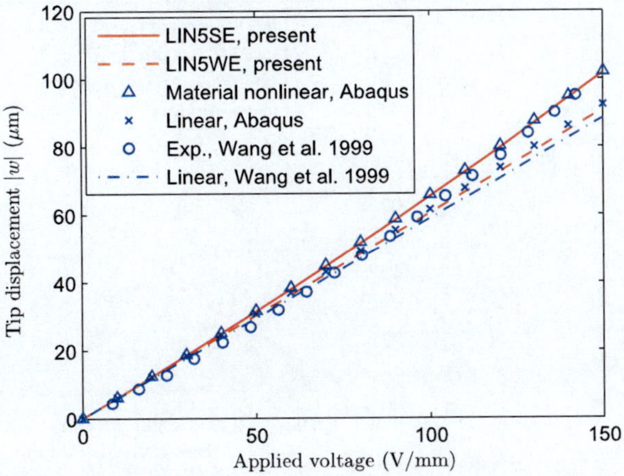

Fig. 6.34 Tip displacement versus electric field for the cantilevered bimorph beam, reprinted from Ref. [23], copyright 2017, with permission from ELSEVIER

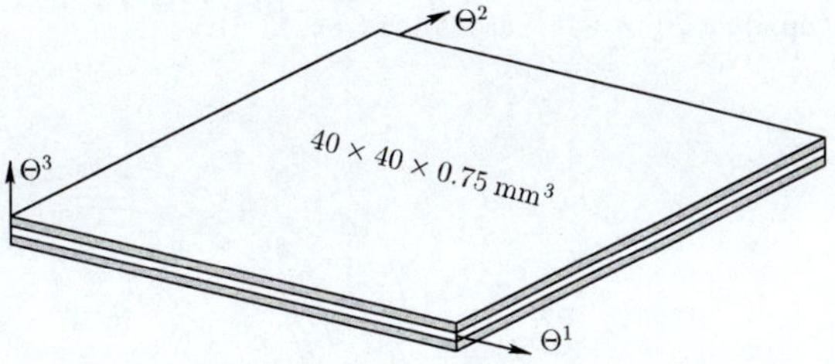

Fig. 6.35 Simply supported piezoelectric plate

of the plate are $40 \times 40 \times 0.75$ mm^3, with equal thickness for each substrate layer (0.25 mm for each layer). The material properties of aluminum and G-1195 are given in Table 6.7. The electrodes connected to the host aluminum plate are assumed to be grounded, and an equal potential is applied on the outer surfaces of G-1195. The resulting electric field through the thickness is considered as constant. The central point displacement versus the electric field is shown in Fig. 6.36. The results of LIN5SE and LIN5WE models have good agreement with those obtained by Abaqus using the equivalent moments applied at corresponding nodes. Furthermore, the figure shows that big gaps happen when strong electric field is applied, for example over 150 V/mm.

6.4.2 *Piezolaminated Semicircular Shell*

Regarding structures undergoing large rotations and under strong electric driving fields, both geometrically and materially nonlinear phenomena have to be considered. In this simulation, a clamped piezolaminated semicircular cylindrical shell is investigated for geometrically nonlinear and electroelastic materially nonlinear analysis, as shown in Fig. 6.37. The semicircular shell is composed of composite

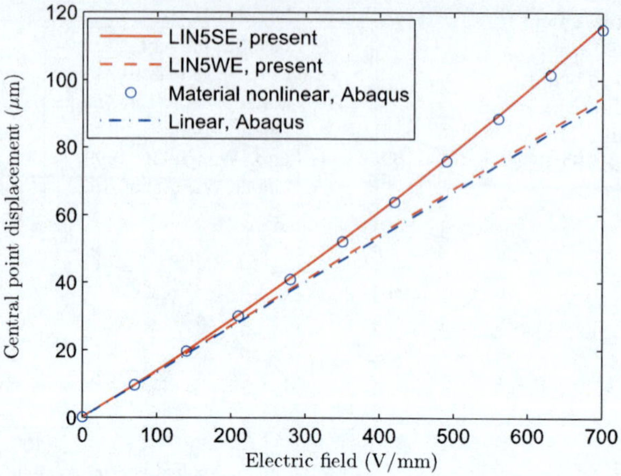

Fig. 6.36 Central point displacement of the simply supported plate, reprinted from Ref. [23], copyright 2017, with permission from ELSEVIER

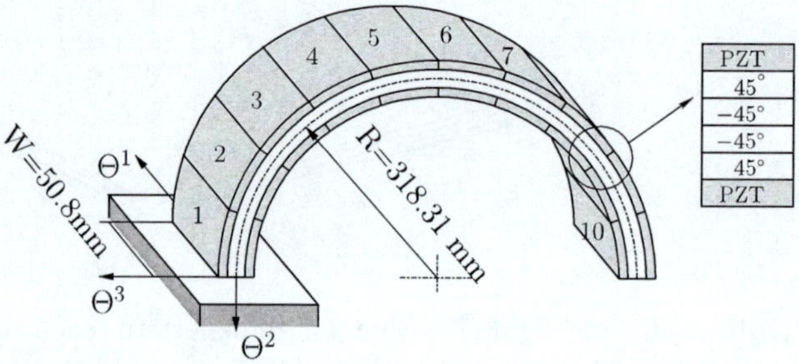

Fig. 6.37 Clamped piezolaminated semicircular cylindrical shell

lamination as host structure and piezoelectric layers bonded on both sides. This example is refined from the Refs. [6, 20]. The host structure is made of graphite/epoxy (T300/976), and the piezoelectric material is chosen as 3203HD. The composite structure has a symmetric lamina sequence of $[\mathrm{P}/45^\circ/-45^\circ]_\mathrm{S}$, where P stands for piezoelectric sublayer. The dimensions of the semicircular are the width 50.8 mm, the radius 318.31 mm and the thickness 1.524 mm. The thickness of each substrate layer is 0.254 mm, including piezoelectric layer. The material properties are given in Table 6.7. The semicircular shell is meshed by 1×10 elements in the axial and hoop directions, respectively, by means of which the convergence was tested by Sze and Yao [21].

An equal electric voltage is imposed on the top and bottom piezoelectric layers. The tip displacements in hoop and radial directions are calculated with accounting for only geometric nonlinearities, as the results shown in Fig. 6.38. Because of the geometric nonlinearity, the linear and nonlinear load-displacement curves have large differences. From the figure, it can be seen that load-displacement curve obtained

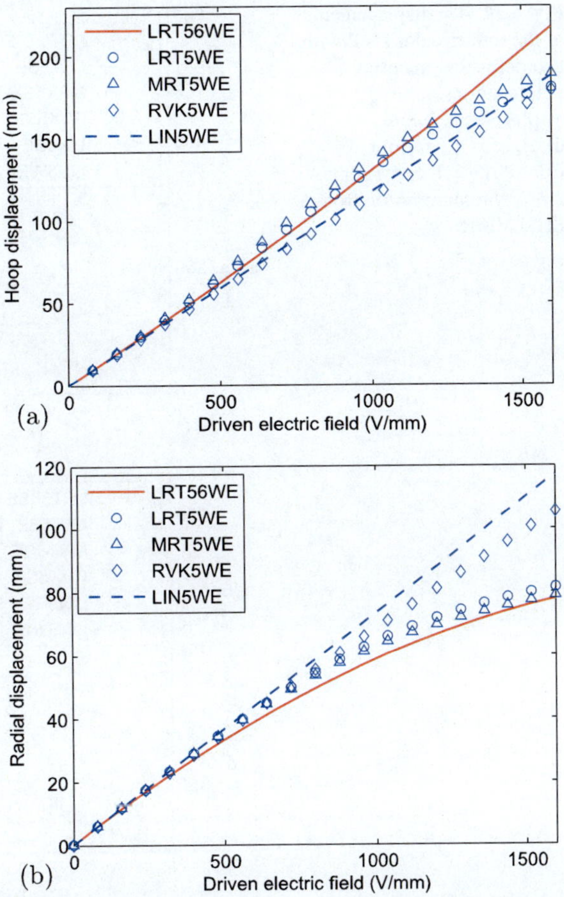

Fig. 6.38 Tip displacements of the semicircular shell with only geometric nonlinearity: **a** hoop displacement, **b** radial displacement, reprinted from Ref. [23], copyright 2017, with permission from ELSEVIER

by the refined von Kármán type nonlinear model (RVK5WE) are close to that of LIN5WE model. The results of LRT56WE model have the largest deviations from those of LIN5WE model. Due to the assumption of moderate rotations, the load-displacement curves of LRT5WE and MRT5WE are very similar, which deviate from those of LRT56WE. From the results, it can clearly show that the semicircular shell with the current boundary and loading conditions are a typical demonstration of large rotation problem. Therefore, the simplified nonlinear models, including RVK5WE, MRT5WE and LRT5WE, fail to predict the structural behavior. For more details of geometrically nonlinear analysis and discussions, it refers to the publications [6, 19, 26].

With consideration of both geometric and material nonlinearities, the hoop and radial displacements of the semicircular shell under the same loading and boundary conditions are presented in Fig. 6.39. Compared to the results of only geometrically nonlinear models the load-displacement curves change dramatically. When

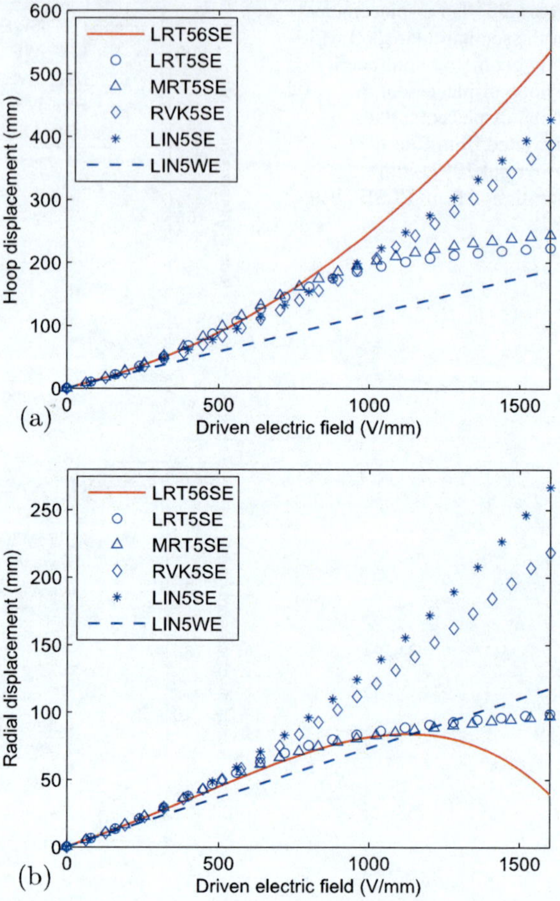

Fig. 6.39 Tip displacements of the semicircular shell with geometric and material nonlinearities: **a** hoop displacement, **b** radial displacement, reprinted from Ref. [23], copyright 2017, with permission from ELSEVIER

low driving electric field is applied (lower than about 500 V/mm), the semicircular shell undergoes only small displacements and rotations. This will lead to weak geometrically nonlinear effect in the shell structure. Therefore, load-deflection curves of all models are close to each other, but are different to linear curve. This can be explained by the fact that electroelastic material nonlinearity dominates the nonlinear structural behavior rather than weak geometrically nonlinear effect.

Increasing the driving electric field, the structure undergoes large displacements and rotations. The influence of geometrically nonlinear effect increases, as the figure shows large deviations occurring between linear and various nonlinear predictions. The RVK5SE model, which includes the simplest nonlinear terms of the strain-displacement relations, has a similar performance of LIN5SE model. The electroelastic nonlinearity is the main nonlinear effect in the structure. Considering more nonlinear strain-displacement terms in the limitation of moderate rotations, MRT5SE and LRT5SE models perform excellently at a relative small electric field (<800 V/mm in this example). For extremely large electric field (>800 V/mm), the LRT56SE curves deviate dramatically from other curves, especially for the radial displacements. This is because the structure undergoes large displacements and rotations. LRT56SE model considers both the electroelastic nonlinearity and full geometric nonlinearity with large rotations.

6.5 Summary

This chapter dealt with nonlinear simulations of composite and smart structures. The present geometrically nonlinear models, RVK5, MRT5, LRT5 and LRT56, were first validated by static analysis and buckling analysis of composite laminated thin-walled structures. Later, the geometrically nonlinear models were applied to numerical analysis of piezolaminated beams, plates and shells. In the final part, the simulation of both geometrically and electroelastic materially nonlinear phenomena were performed on piezoelectric integrated smart structures.

References

1. C.T. Sun, H. Chin, Analysis of asymmetric composite laminates. AIAA J. **26**, 714–718 (1988)
2. J.N. Reddy, On refined computational models of composite laminates. Int. J. Numer. Methods Eng. **27**, 361–382 (1989)
3. Y. Basar, Y. Ding, R. Schultz, Refined shear-deformation models for composite laminates with finite rotations. Int. J. Solids Struct. **30**, 2611–2638 (1993)
4. I. Kreja, R. Schmidt, Large rotations in first-order shear deformation FE analysis of laminated shells. Int. J. Non-Linear Mech. **41**, 101–123 (2006)
5. S.Q. Zhang, R. Schmidt, Large rotation theory for static analysis of composite and piezoelectric laminated thin-walled structures. Thin-Walled Struct. **78**, 16–25 (2014)
6. S.Q. Zhang, R. Schmidt, Static and dynamic FE analysis of piezoelectric integrated thin-walled composite structures with large rotations. Compos. Struct. **112**, 345–357 (2014)
7. N. Stander, A. Matzenmiller, E. Ramm, An assessment of assumed strain methods in finite rotation shell analysis. Eng. Comput. **6**, 58–66 (1989)
8. A.F. Saleeb, T.Y. Chang, W. Graf, S. Yingyeunyong, A hybrid/mixed model for non-linear shell analysis and its applications to large-rotation problems. Int. J. Numer. Methods Eng. **29**, 407–446 (1990)
9. C. Sansour, H. Bufler, An exact finite rotation shell theory, its mixed variational formulation and its finite element implementation. Int. J. Numer. Methods Eng. **34**, 73–115 (1992)
10. L. Jiang, M.W. Chernuka, A simple four-noded corotational shell element for arbitrarily large rotations. Comput. Struct. **53**, 1123–1132 (1994)
11. A. Masud, C.L. Tham, W.K. Liu, A stabilized 3-D co-rotational formulation for geometrically nonlinear analysis of multi-layered composite shells. Comput. Mech. **26**, 1–12 (2000)
12. K.Y. Sze, X.H. Liu, S.H. Lo, Popular benchmark problems for geometric nonlinear analysis of shells. Finite Elem. Anal. Des. **40**, 1551–1569 (2004)
13. R.A. Arciniega, J.N. Reddy, Tensor-based finite element formulation for geometrically nonlinear analysis of shell structures. Comput. Methods Appl. Mech. Eng. **196**, 1048–1073 (2007)
14. S. Saigal, R.K. Kapania, T.Y. Yang, Geometrically nonlinear finite element analysis of imperfect laminated shells. J. Compos. Mater. **20**, 197–214 (1986)

15. G. Laschet, J.P. Jeusette, Postbuckling finite element analysis of composite panels. Compos. Struct. **14**, 35–48 (1990)
16. B. Brank, D. Perić, B. Damjanić, On implementation of a nonlinear four node shell finite element for thin multilayered elastic shells. Comput. Mech. **16**, 341–359 (1995)
17. S. Yi, S.F. Ling, M. Ying, Large deformation finite element analyses of composite structures integrated with piezoelectric sensors and actuators. Finite Elem. Anal. Des. **35**, 1–15 (2000)
18. S. Lentzen, R. Schmidt, A geometrically nonlinear finite element for transient analysis of piezolaminated shells, in *Proceedings Fifth EUROMECH Nonlinear Dynamics Conference* (Eindhoven, Netherlands, 7–12 August 2005), pp. 2492–2500
19. S.Q. Zhang, R. Schmidt, Large rotation FE transient analysis of piezolaminated thin-walled smart structures. Smart Mater. Struct. **22**, 105025 (2013)
20. H.S. Tzou, R. Ye, Analysis of piezoelastic structures with laminated piezoelectric triangle shell elements. AIAA J. **34**, 110–115 (1996)
21. K.Y. Sze, L.Q. Yao, Modeling smart structures with segmented piezoelectric sensors and actuators. J. Sound Vib. **235**, 495–520 (2000)
22. Q.M. Wang, Q. Zhang, B. Xu, R. Liu, L.E. Cross, Nonlinear piezoelectric behavior of ceramic bending mode actuators under strong electric fields. J. Appl. Phys. **86**(6), 3352–3360 (1999)
23. S.Q. Zhang, G.Z. Zhao, S.Y. Zhang, R. Schmidt, X.S. Qin, Geometrically nonlinear FE analysis of piezoelectric laminated composite structures under strong driving electric field. Compos. Struct. **181**, 112–120 (2017)
24. S. Kapuria, M.Y. Yasin, A nonlinear efficient layerwise finite element model for smart piezolaminated composites under strong applied electric field, in *Smart Materials and Structures* (2013)
25. L.Q. Yao, J.G. Zhang, L. Lu, M.O. Lai, Nonlinear extension and bending of piezoelectric laminated plate under large applied field actuation. Smart Mater. Struct. **13**, 404–414 (2004)
26. S.Q. Zhang, Z.X. Wang, X.S. Qin, G.Z. Zhao, R. Schmidt, Geometrically nonlinear analysis of composite laminated structures with multiple macro-fiber composite (MFC) actuators. Compos. Struct. **150**, 62–72 (2016)

Chapter 7
Numerical Analysis of Macro-fiber Composite Structures

Abstract This chapter deals with the simulation of MFC bonded structures using the numerical models developed in previous chapters. First the model is validated by MFC bonded plate structure, in which two typical types of MFCs are considered, MFC-d31 and MFC-d33. Then, various geometrically nonlinear models are applied to compute multi-MFC integrated pate and cylindrical shell structures. In both linear and nonlinear analysis, various piezo-fiber orientation angles are considered, to demonstrate the influences on structural response.

7.1 Linear Analysis of MFC Structures

7.1.1 Validation Test

The first validation test is conducted on a cantilevered plate bonded with MFC-d33 patches on the top and bottom surfaces, which was proposed by Bowen et al. [1], as shown in Fig. 7.1. The host structure is an aluminum plate, with the material properties of the Young's modulus Y = 70 GPa and the Possion's ratio $\nu = 0.32$. The dimensions of the MFC plate are $300 \times 75 \times 1.97$ mm^3, while those of MFC-d33 patches are $85 \times 57 \times 0.3$ mm^3 (M8557-P1, Smart Material Corp. [2]). The MFC patches are bonded at a distance $d = 15$ mm away from the clamped edge, see Fig. 7.1. The material parameters of MFC-d33 are given in Table 7.1, based on the studies of Williams et al. [3] and Bowen et al. [1], which are slightly different from those provided by Smart Material Corp. [2].

A constant voltage loading of 400 V (electric field 400/0.5 V/mm) is applied on the top MFC-d33 patch. The vertical displacements of the central line are calculated and presented in Fig. 7.2, with the corresponding date listed in Table 7.2. With the comparison of the current results and those obtained by ANSYS and experimental investigations in Bowen et al. [1], a good agreement has been reached. Thus, the present FE model is verified to be accurate enough for the simulation of MFC actuated structures (Fig. 7.2 and Table 7.2).

S.-Q. Zhang, *Nonlinear Analysis of Thin-Walled Smart Structures*, Springer Tracts in Mechanical Engineering, https://doi.org/10.1007/978-981-15-9857-9_7

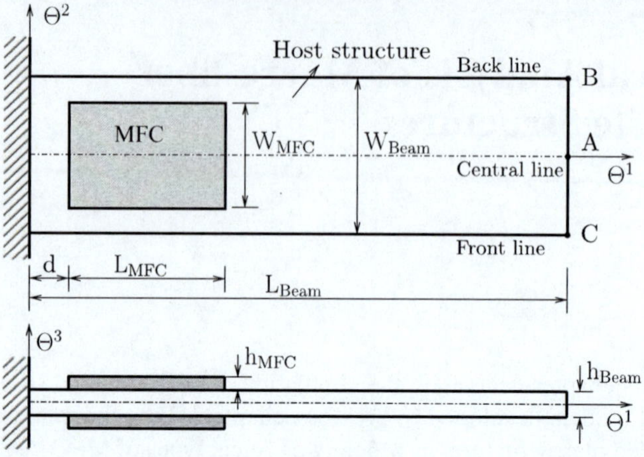

Fig. 7.1 Schematic figure of the MFC bonded smart plate

Table 7.1 Material properties

MFC-d33	MFC-d31	T300/976
$\tilde{Y}_3 = 29.4$ GPa	$\tilde{Y}_1 = 30.336$ GPa	$\tilde{Y}_1 = 150$ GPa
$\tilde{Y}_2 = \tilde{Y}_1 = 15.2$ GPa	$\tilde{Y}_2 = \tilde{Y}_3 = 15.857$ GPa	$\tilde{Y}_2 = \tilde{Y}_3 = 9$ GPa
$\tilde{G}_{32} = \tilde{G}_{31} = 6.06$ GPa	$\tilde{G}_{12} = \tilde{G}_{13} = 5.515$ GPa	$\tilde{G}_{12} = \tilde{G}_{13} = 7.1$ GPa
$\tilde{G}_{21} = 5.79$ GPa	$\tilde{G}_{23} = 5.515$ GPa	$\tilde{G}_{23} = 2.5$ GPa
$\tilde{\nu}_{32} = \tilde{\nu}_{31} = 0.312$	$\tilde{\nu}_{12} = \tilde{\nu}_{13} = 0.31$	$\tilde{\nu}_{12} = \tilde{\nu}_{13} = 0.3$
$\tilde{\nu}_{21} = 0.312$	$\tilde{\nu}_{23} = 0.438$	$\tilde{\nu}_{23} = 0.3$
$\tilde{d}_{33} = 467 \times 10^{-12}$ m/V	$\tilde{d}_{31} = -170 \times 10^{-12}$ m/V	
$\tilde{d}_{32} = -210 \times 10^{-12}$ m/V	$\tilde{d}_{32} = -100 \times 10^{-12}$ m/V	
$h_E = 0.5$ mm	$h_E = 0.18$ mm	

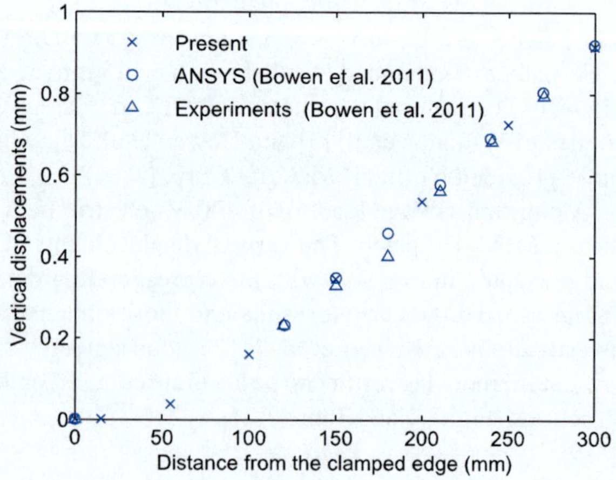

Fig. 7.2 Central line deflection of the MFC-d33 bonded plate for validation test, reprinted from Ref. [4], copyright 2015, with permission from ELSEVIER

Table 7.2 Numerical values for the present result in Fig. 7.2, reprinted from Ref. [4], copyright 2015, with permission from ELSEVIER

Θ^1 (mm)	Deflection (mm)
0	0
15	6.1445×10^{-4}
55	0.0372
100	0.1593
150	0.3453
200	0.5344
250	0.7242
300	0.9141

7.1.2 Isotropic Plate Bonded with MFC-d31 Patches

A similar clamped plate boned with MFC-d31 on the top and bottom surfaces is considered in this simulation. The host plate is made of aluminum and the piezoelectric patch is chosen as MFC-d31 type (M8528-P3, Smart Material Corp. [2]). The material parameters of aluminum are the same with the example in validation test, and those of MFC-d31 are listed in Table 7.1. The dimensions for the host structure are $300 \times 75 \times 2$ mm^3, and those for the MFC-d31 patch are $85 \times 28 \times 0.3$ mm^3. The MFC-d31 patches are bonded at a distance of $d = 15$ from the clamped edge.

The top MFC-d31 patch of the smart plate is subjected to a driving voltage of 200 V, equivalent to the electric field 200/0.18 V/mm. The central line deflections are computed and plotted in Fig. 7.3, with numerical values listed in Table 7.3. From the figure, the deflection curve comprises three parts, namely, a straight line (from 0 to 15 mm in Θ^1-axis direction), a curved line (from 15 mm to 100 mm, where the MFC-d31 patches are bonded), and another straight line (from 100 mm to 300 mm). The figure shows that the central line deflections of MFC-d31 plate are much smaller than those of MFC-d33 plate. The reasons are first the dimensions of MFC-d31 patch is smaller, second piezoelectric coefficient of MFC-d31 is only about 36% of MFC-d33.

7.1.3 Isotropic Plate with MFC-d33 Patches Having Arbitrary Fiber Orientation

In this example, a clamped MFC plate is again considered for investigation of the influence of piezo fiber orientations, as shown in Fig. 7.1. The dimensions of the host aluminum plate and MFC-d33 patches are respectively $300 \times 75 \times 2$ mm^3 and $85 \times 57 \times 0.3$ mm^3. The material properties of MFC-d33 patches are given in Table 7.1. An electric driving voltage of 400 V is applied on the top layer of MFC, and the bottom layer is short circuited. With consideration of different piezo fiber orientation, the vertical displacement of point A and the twist of the plate (defined as $w_B - w_C$) are

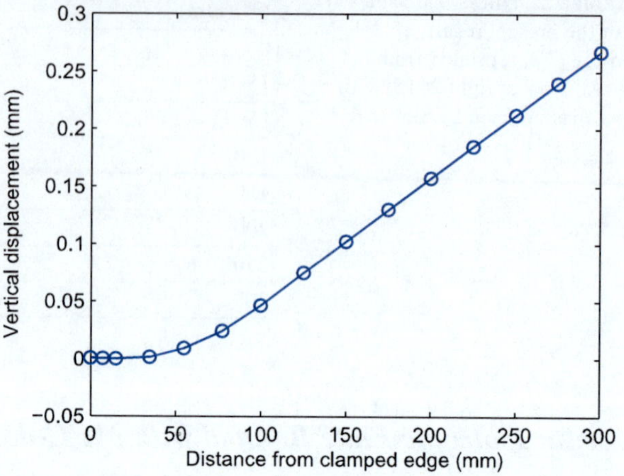

Fig. 7.3 Central line deflection of the aluminum plate bonded with MFC-d31 patches, reprinted from Ref. [4], copyright 2015, with permission from ELSEVIER

Table 7.3 Numerical values for the curve in Fig. 7.3, reprinted from Ref. [4], copyright 2015, with permission from ELSEVIER

Θ^1 (mm)	Deflection (mm)
0	0
15	-5.8704×10^{-4}
55	0.0089
100	0.0462
150	0.1020
200	0.1571
250	0.2122
300	0.2672

obtained and illustrated in Fig. 7.4. The corresponding numerical values are listed in Table 7.4.

From the figure, it can be seen that the vertical displacements of point A decrease to zeros as the piezo fiber angle increase to about 60°, then the displacements go into negative. For the twist, it first increases from zero to the maximum value at the fiber angle of 45°, then reduces to zero as the material angle of MFC-d33 increases to 90°. The displacements turn from positive to negative results from the sign of the main piezoelectric coefficient changes from positive value of d_{33} to the negative one d_{31}. In Table 7.4, it shows that the twist is almost equal to zero when the fiber angle of MFC-d33 is 0° or 90°. Theoretically, the twists should be zero at the angle of 0° or 90°. The small deviations are resulting from the error margin of the numerical computations.

To deep illustrate the deformation of MFC plate, the deformed shapes are plotted in 3-dimensional view for the piezoelectric fiber reinforcement aligning in the angles of 0°, 30°, 45°, 60°, 75° and 90°, as shown in Fig. 7.5.

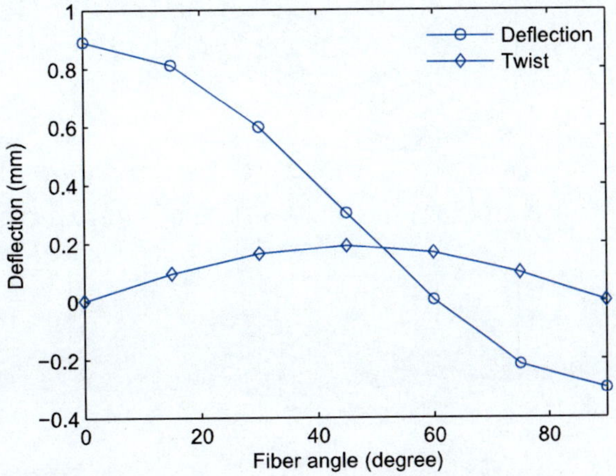

Fig. 7.4 Vertical deflections and twist of the aluminum plate with MFC-d33 patches, reprinted from Ref. [4], copyright 2015, with permission from ELSEVIER

Table 7.4 Numerical values for the vertical deflections and twist (mm), reprinted from Ref. [4], copyright 2015, with permission from ELSEVIER

Piezo fiber angle	Deflection (w_A)	Twist (w_B–w_C)
0°	0.8897	7.3074×10^{-11}
15°	0.8114	0.0935
30°	0.5968	0.1624
45°	0.3019	0.1886
60°	3.3226×10^{-3}	0.1644
75°	−0.2182	0.0955
90°	−0.3001	2.1209×10^{-11}

Additionally, the vertical displacements of the front line, central line and back line are illustrated in Fig. 7.6, where the plate twist can be recognized clearly. The results show that the vertical displacements of the front, central and back lines are identical in the case of the fiber orientation angle of MFC is 0° or 90°. Because there is no twist occurring in such configuration. From Fig. 7.6a, it can be observed that a slight difference existing between the displacements of the central and front (or back) lines at the bonding area. This is because the MFC elongates not only in the length direction, but also in the width direction. The is discrepancy will be enlarged if the fiber orientation of MFC patches is 90°, as illustrated in Fig. 7.6f. The reason is explained that the coefficient d_{33} is much larger that d_{31}.

To investigate the stresses in the bonding area, the in-plane longitudinal stress ε_{11} and the transverse shear stress ε_{13} are analyzed and presented in Fig. 7.7. The figures indicate that the longitudinal stresses are very strong in the case of fiber

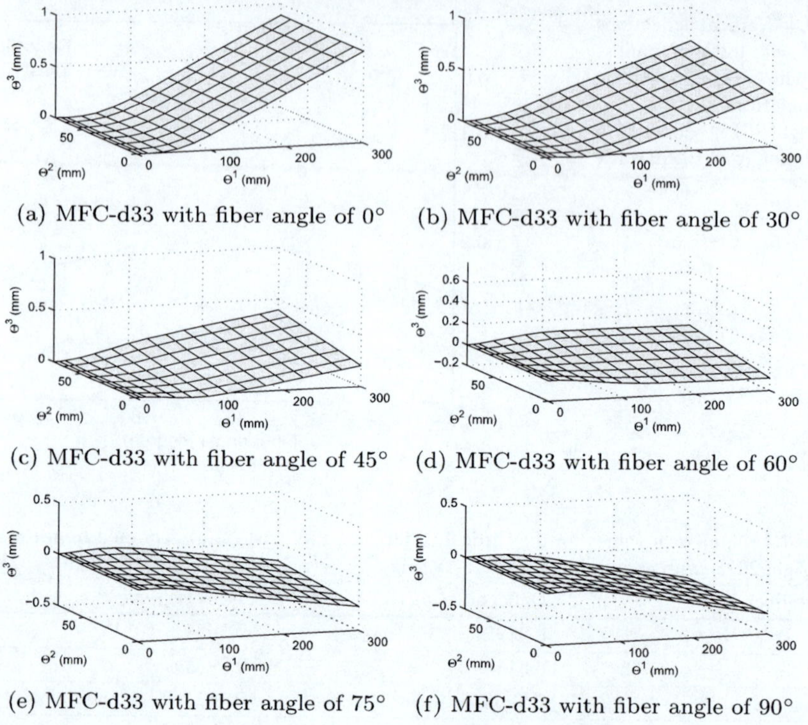

(a) MFC-d33 with fiber angle of 0°
(b) MFC-d33 with fiber angle of 30°
(c) MFC-d33 with fiber angle of 45°
(d) MFC-d33 with fiber angle of 60°
(e) MFC-d33 with fiber angle of 75°
(f) MFC-d33 with fiber angle of 90°

Fig. 7.5 Surface shapes of the aluminum plate with MFC-d33 patches having different fiber angles, reprinted from Ref. [4], copyright 2015, with permission from ELSEVIER

orientation angle of 0°, 30° and 45°. Additionally, the transverse shear stresses show a complicated behavior with positive and negative stagger arrangement in the case of piezo fiber orientation angle of 30°, 45° and 60°.

7.1.4 Composite Plate with MFC-d33 Patches Having Arbitrary Fiber Orientation

In this simulation, a very similar MFC plate is studied as shown in Fig. 7.1, in which the host structure is considered as a composite laminated structure. The plate is made of T300/976, stacked symmetrically as $[90/0]_s$. The composite laminated plate is bonded with two MFC-d33 patches on the top and bottom surfaces. The dimensions of the host plate are $300 \times 75 \times 2$ mm^3 and those of the MFC-d33 patches are $85 \times 28 \times 0.3$ mm^3. The thickness for each composite sublayer is 0.5 mm. The material parameters for T300/976 and MFC-d33 can be found in Table 7.1. An electric driving voltage of 400 V is applied on the top MFC patch. The 3-dimensional

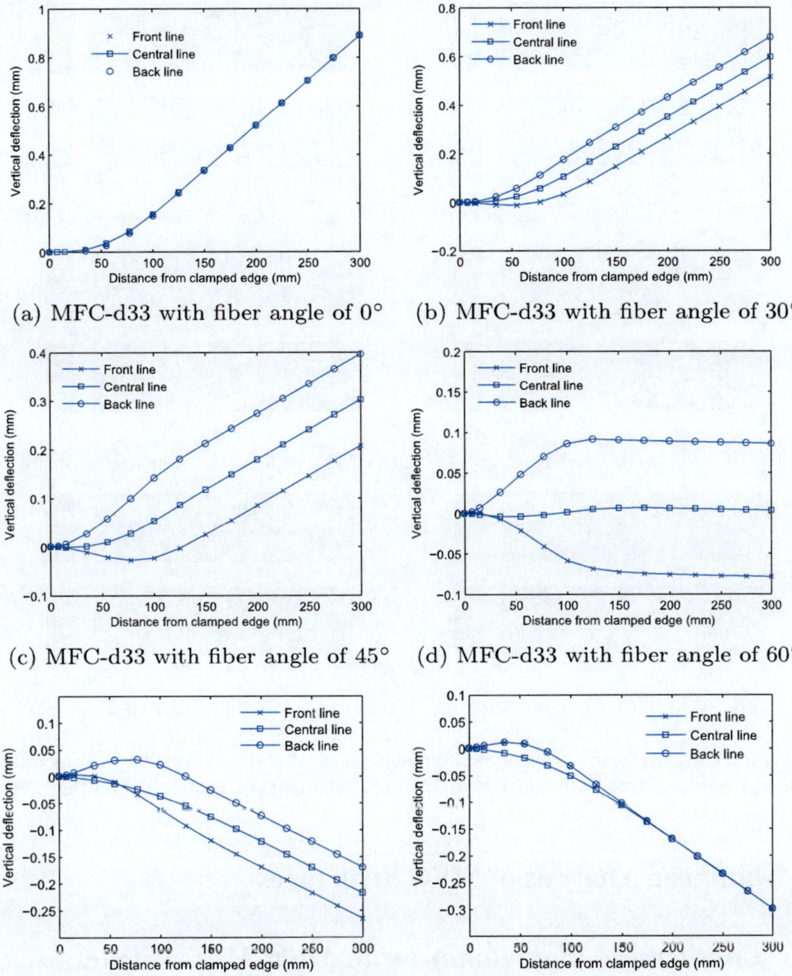

(a) MFC-d33 with fiber angle of 0°
(b) MFC-d33 with fiber angle of 30°
(c) MFC-d33 with fiber angle of 45°
(d) MFC-d33 with fiber angle of 60°
(e) MFC-d33 with fiber angle of 75°
(f) MFC-d33 with fiber angle of 90°

Fig. 7.6 Line shapes of the aluminum plate with MFC-d33 patches having different fiber angles, reprinted from Ref. [4], copyright 2015, with permission from ELSEVIER

deformation shapes are plotted with consideration of various fiber orientation angles, as the results shown in Fig. 7.8. The corresponding vertical displacements of the front, central and back lines are illustrated in Fig. 7.9.

An analogous conclusion with the MFC aluminum plate can be drawn by observation of the results. Due to the symmetrical stacking sequence of the composite lamina, there is no twist occurring in the case of piezo fiber angle of 0° or 90°. However, a big twist deflections is indicated by the 3-dimensional deformation shapes when the piezo fiber angle equals to 30° and 45°.

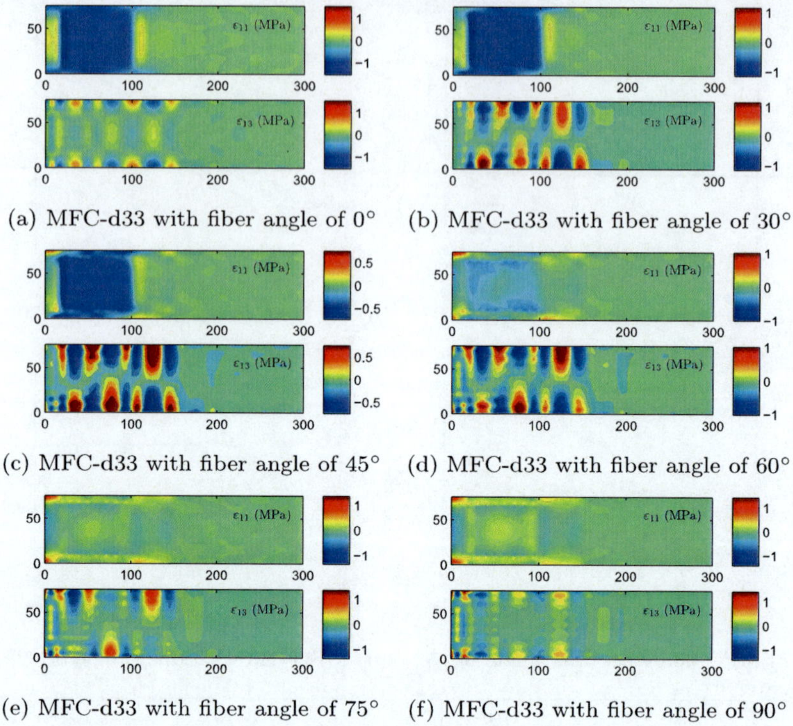

Fig. 7.7 Stress (ε_{11},ε_{13}) distribution of the aluminum plate with MFC-d33 patches having different fiber angles, reprinted from Ref. [4], copyright 2015, with permission from ELSEVIER

7.2 Nonlinear Analysis of MFC Structures

7.2.1 Cantilevered Plate Bonded with Multi-MFC Patches

In this section, geometrically nonlinear analysis are performed on multi-MFC bonded structures. First, a clamped plate integrated with multi-patches of MFC-d33 is studied, as shown in Fig. 7.10. The host plate structure is made of T300/976 graphite/epoxy angle-ply laminates, with the stacking sequence of $[-45°/45°]$ from the lower to the upper positions. The dimensions of the plate and MFC-d33 patches are respectively $284 \times 44 \times 0.2$ mm^3 and $56 \times 28 \times 0.3$ mm^3. The thickness for each substrate layer of composites is 0.1 mm. The distribution spacing of MFC patches on the plate is $d = 12$ mm. The material properties are give in Table 7.1.

Four different discretization schemes are investigated in order to study the convergence performance, namely 9×3, 13×3, 13×4 and 17×4, as shown in Fig. 7.11. In the simulation, eight-node quadrilateral elements with five mechanical DOFs at each node and one electric DOF on each element are considered using uniformly reduced integration scheme. An equal driving electric voltage of 800 V

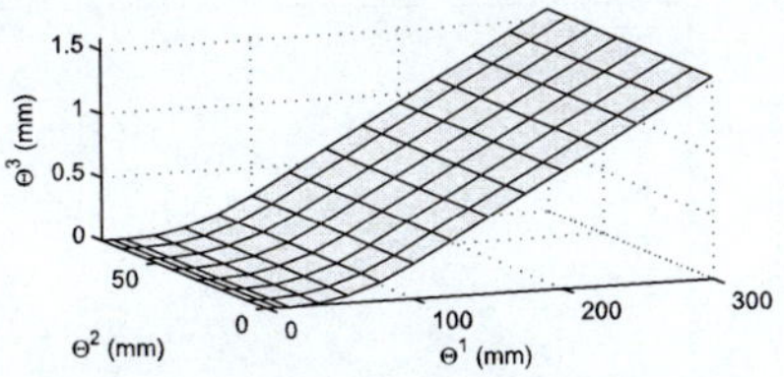

(a) MFC-d33 with fiber angle of 0°

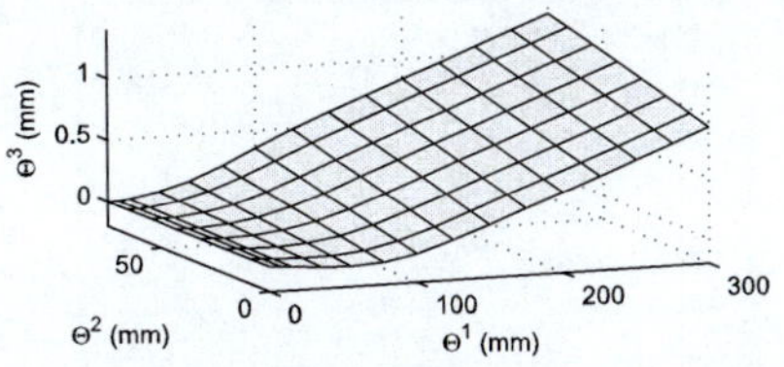

(b) MFC-d33 with fiber angle of 30°

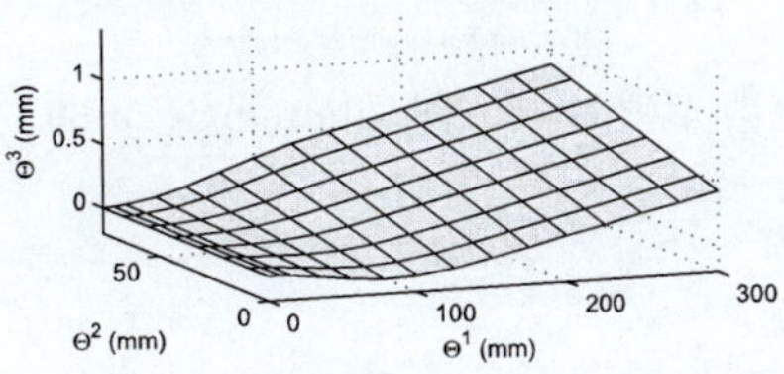

(c) MFC-d33 with fiber angle of 45°

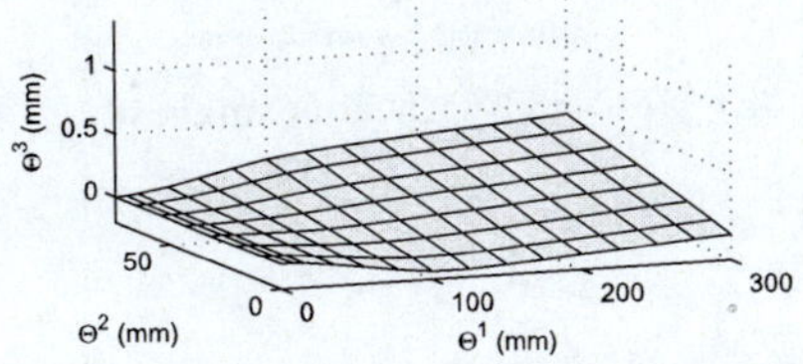

(d) MFC-d33 with fiber angle of 60°

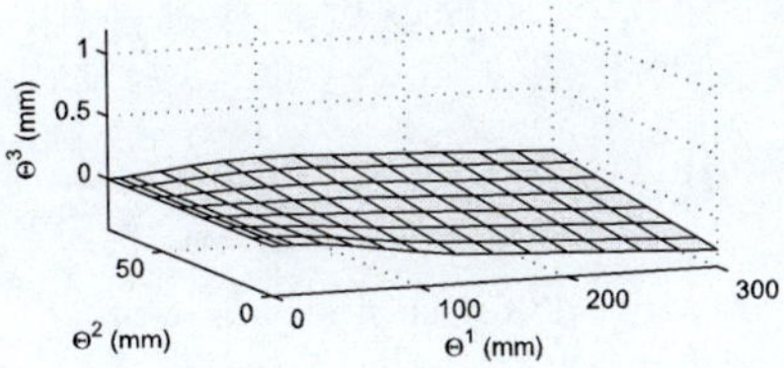

(e) MFC-d33 with fiber angle of 75°

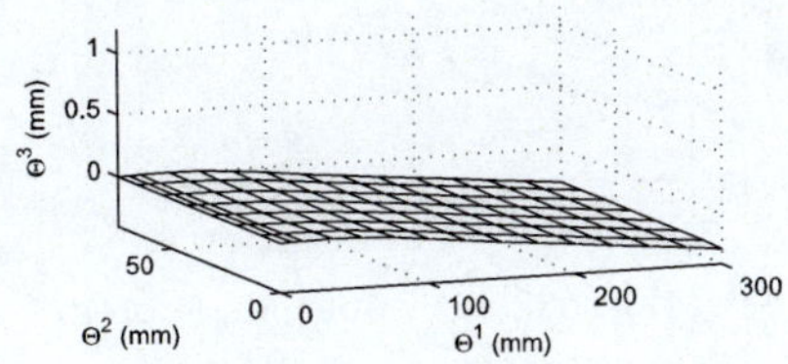

(f) MFC-d33 with fiber angle of 90°

Fig. 7.8 Surface shapes of the composite plate with MFC-d33 patches having different fiber angles, reprinted from Ref. [4], copyright 2015, with permission from ELSEVIER

($E_1 = 800/0.5$ V/mm) is imposed on all MFC patches. The tip displacements of the multi-MFC bonded plate are computed and presented in Table 7.5. From the results, it implies that all the discretization schemes are converged due to the extremely small deviation between all the cases. In the later analysis, the mesh of 17×4 is employed.

Considering different fiber orientation angles of MFC patches, the vertical displacements w_{11} at node 11 and the twist $|w_{10}\text{-}w_{12}|$ are obtained by using various geometrically nonlinear models, which are illustrated in Figs. 7.12 and 7.13, respectively.

From Fig. 7.12, it indicates that the largest displacement always occurs at the piezo fiber orientation angle of 0° for all linear and nonlinear models. As discussed in the above chapters, the performances of LRT5 and MRT5 models are very similar, leading to the results with small deviation. In this example, the RVK5 model predicts results close to LRT56 model. This does not illustrate that RVK5 either has higher accuracy than LRT5 and MRT5, nor as same as LRT56. The figure shows large discrepancies existing between the results of LIN5, RVK5, MRT5, and LRT5, which implies that the structure has occurred large rotations. Therefore, LRT56 model should be used for the correct static displacement response. The results of twist

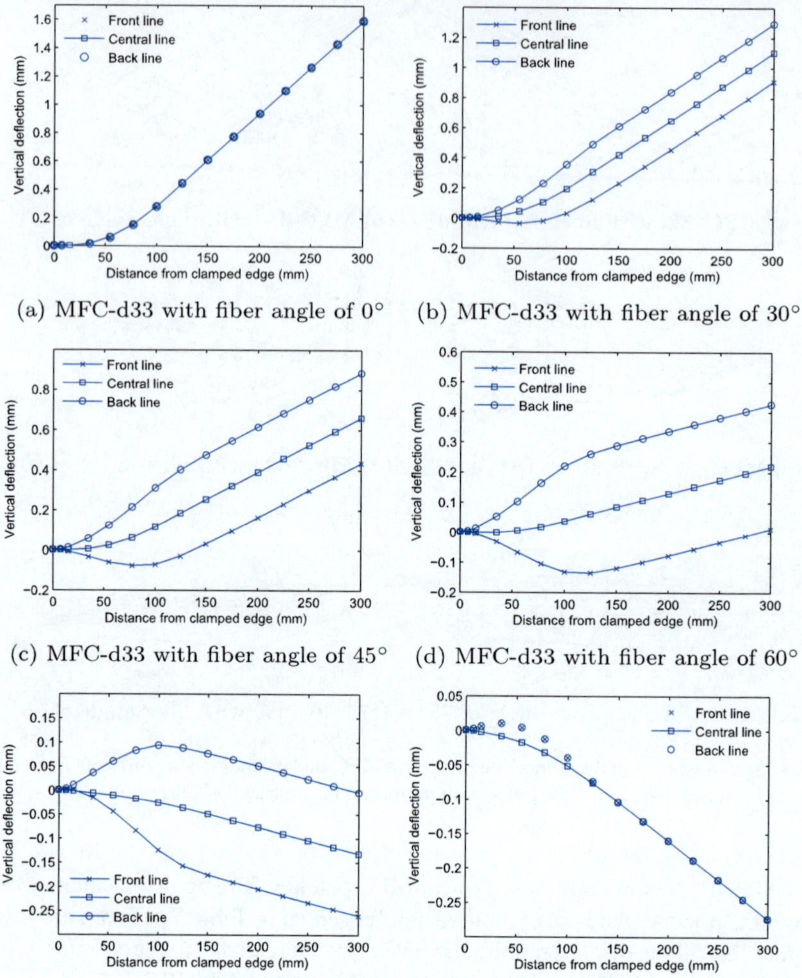

(a) MFC-d33 with fiber angle of 0°
(b) MFC-d33 with fiber angle of 30°
(c) MFC-d33 with fiber angle of 45°
(d) MFC-d33 with fiber angle of 60°
(e) MFC-d33 with fiber angle of 75°
(f) MFC-d33 with fiber angle of 90°

Fig. 7.9 Line shapes of the composite plate with MFC-d33 patches having different fiber angles, reprinted from Ref. [4], copyright 2015, with permission from ELSEVIER

in Fig. 7.13 show that the largest twists occur when the piezo fiber orientation angle is 45° for all different linear and nonlinear models. The relations between each result has a same tendency with the relations in the load-displacement figure.

To investigate deeply on the MFC driving performance, the deformation shapes of multi-MFC plate are studies under the same boundary and loading conditions. Using LIN5 and LRT56 models, the linear and nonlinear deformations for various piezo fiber orientation angles are observed and presented in Fig. 7.14. At each subfigure, the upper group of plots are the deformed shape in 3-dimensional space, while the

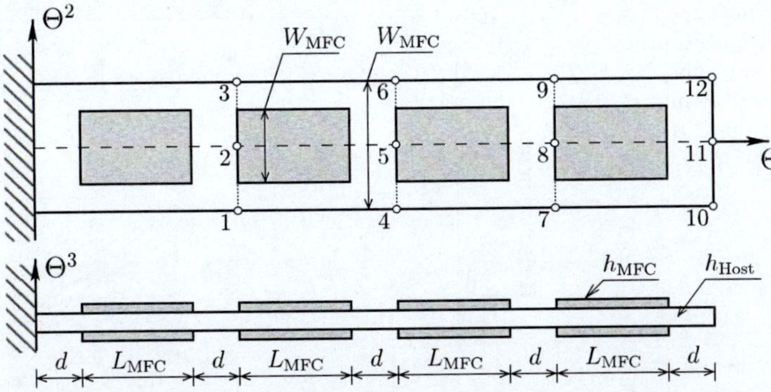

Fig. 7.10 Cantilevered plate bonded with multiple MFC actuators, reprinted from Ref. [5], copyright 2016, with permission from ELSEVIER

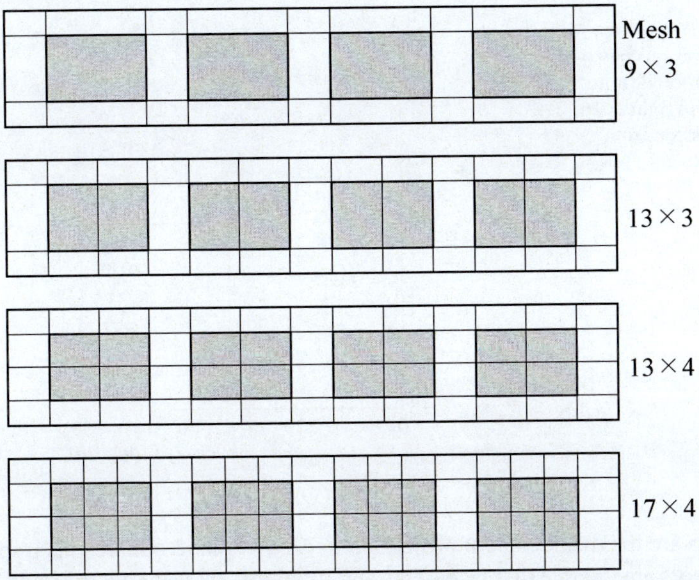

Fig. 7.11 Cantilevered plate bonded with multiple MFC actuators, reprinted from Ref. [5], copyright 2016, with permission from ELSEVIER

Table 7.5 Convergence study

Mesh	Tip displacement (mm)	Deviation (%)
9 × 3	86.0833	0.1120
13 × 3	85.9194	0.3022
13 × 4	86.2189	0.0454
17 × 4	86.1798	0

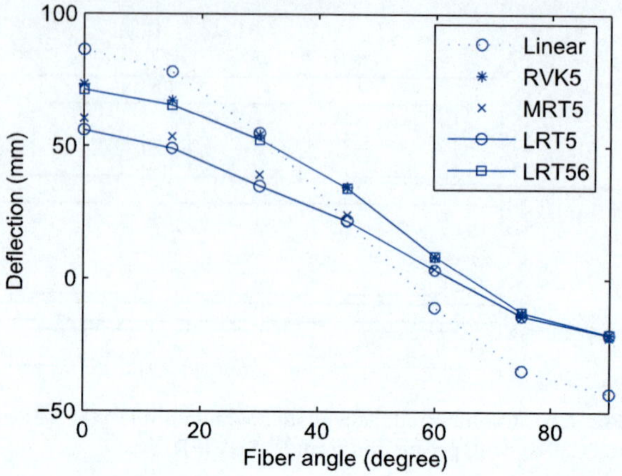

Fig. 7.12 Vertical tip deflection of the MFC plate with various piezo-fiber angles, reprinted from Ref. [5], copyright 2016, with permission from ELSEVIER

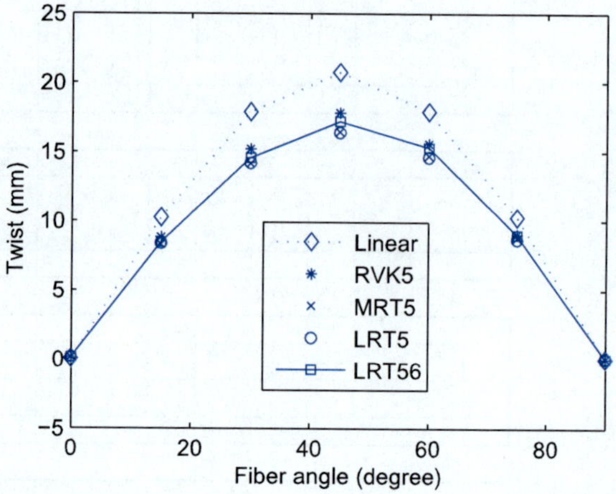

Fig. 7.13 Twist of the MFC plate with various piezo-fiber angles, reprinted from Ref. [5], copyright 2016, with permission from ELSEVIER

lower ones are the displacement projection in Θ^1-Θ^2 plane. The results implies that the structures are in large displacements and rotations, since there are big differences in the static response of LIN5 and LRT56 models. The results also indicate that the twists in the Θ^1-Θ^2 plane are not negligible, especially in the case of MFC piezo fiber orientation angle of 30° and 45°.

7.2.2 Cantilevered Semicircular Cylindrical Shell with Multi-MFC Patches

In this example, a cantilevered semicircular cylindrical shell integrated with multiple MFC patches are studied, as shown in Fig. 7.15. The arc length and width of the mid-

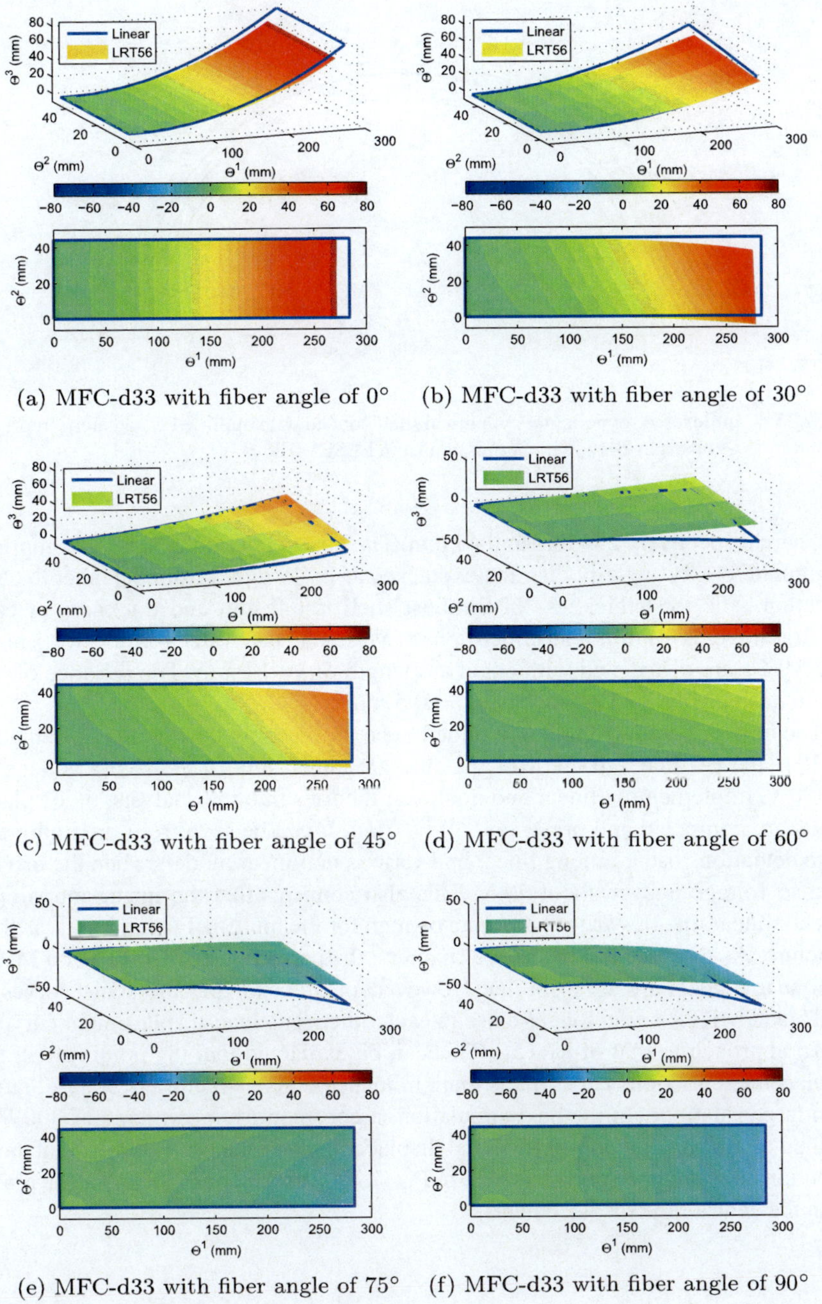

(a) MFC-d33 with fiber angle of 0°
(b) MFC-d33 with fiber angle of 30°
(c) MFC-d33 with fiber angle of 45°
(d) MFC-d33 with fiber angle of 60°
(e) MFC-d33 with fiber angle of 75°
(f) MFC-d33 with fiber angle of 90°

Fig. 7.14 Surface shapes of the composite plate with MFC-d33 patches having different fiber angles, reprinted from Ref. [5], copyright 2016, with permission from ELSEVIER

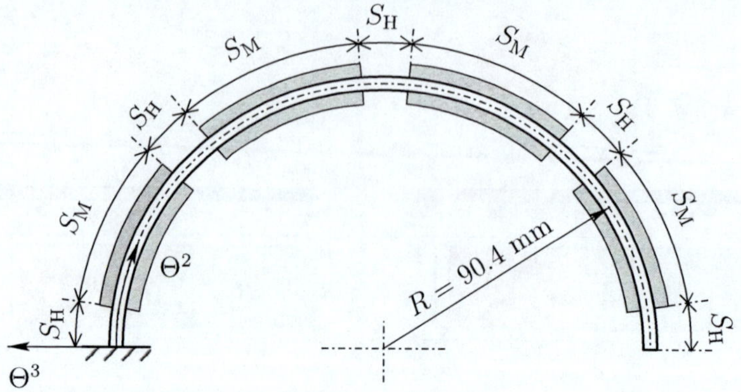

Fig. 7.15 Cantilevered semicircular cylindrical shell bonded with multi-MFC actuators, reprinted from Ref. [5], copyright 2016, with permission from ELSEVIER

plane are respectively 284 mm and 44 mm. The host semicircular shell is comprised of graphite/epoxy angle-ply laminates stacked as [−45°/45°] from the inner to outer direction. The total thickness of the host shell is 0.2 mm and thickness of each substrate layer is 0.1 mm. The dimensions of MFC-d33 patch are 56 × 28 × 0.3 mm^3. The MFCs are distributed with equal arc length $S_{\mathrm{H}} = 12$ mm. The distance of two electrodes in MFC-d33 patches is $h_{\mathrm{E}} = 0.5$ mm.

The fiber orientation angle is assumed to be 0°, meaning that the fiber orientation is along the Θ^1 line. All the MFC patches are under a uniform driving voltage up to 300 V. Implementing linear and nonlinear theories into the analysis, tip displacements are computed and presented in Fig. 7.16. It can be seen from the figure that large deviations occur among linear and various nonlinear models when the driving electric voltage is over about 50 V. This also confirms that the linear and simplified nonlinear theories are not accurate enough for the multi-MFC semicircular shell structure. As discussed in the above chapters, the performance of LRT5 and MRT5 nonlinear models are very similar. However, in this example the static curves of LRT5 and MRT5 are no longer close to each other. The load-displacement curve of LRT5 approach to that of LRT56. This can be explained that the nonlinear strain-displacement relations perform the main influence on the structural response, rather than large rotations. In the next simulation, a constant driving voltage of 300 V is applied to all MFC patches. The radial displacements of the central line in the hoop direction are calculated and presented in Fig. 7.17. Similar conclusions can be drawn from the central line displacements.

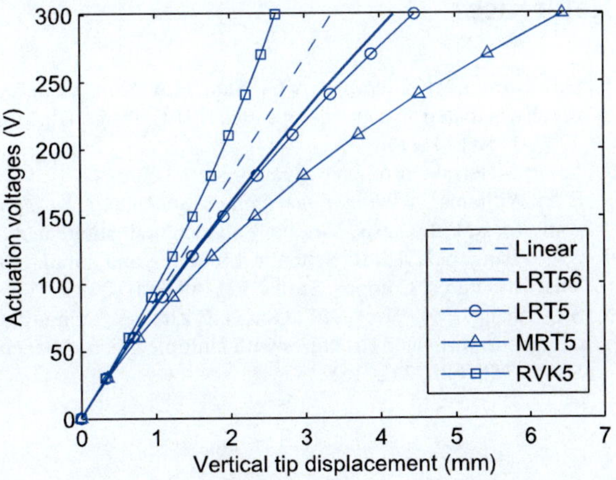

Fig. 7.16 The radial tip displacements under various actuation loads, reprinted from Ref. [5], copyright 2016, with permission from ELSEVIER

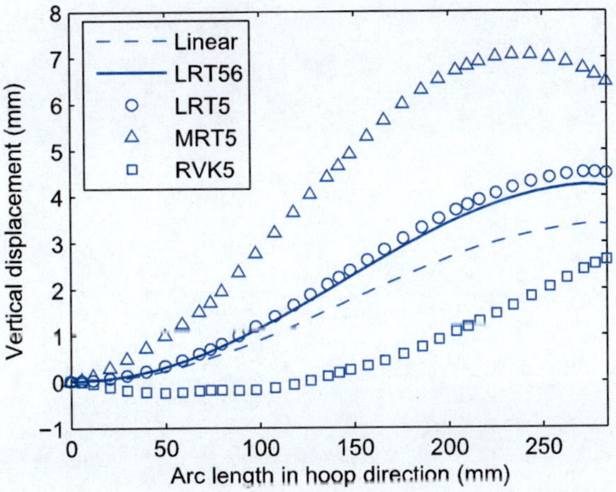

Fig. 7.17 The radial displacements of the central line in the hoop direction, reprinted from Ref. [5], copyright 2016, with permission from ELSEVIER

7.3 Summary

The chapter conducted numerical investigations on MFC bonded smart structures. First, an MFC plate was analyzed for validation test of MFC models. The finite element models of MFC-d31 or MFC-d33 were performed on MFC monolithic or composite plates with consideration of various piezo fiber orientations. In the last part, geometrically nonlinear analysis of multi-MFC bonded plate and semicircular shell was studied.

References

1. C.R. Bowen, P.F. Giddings, A.I.T. Salo, H.A. Kim, Modeling and characterization of piezoelectrically actuated bistable composites. IEEE Trans. Ultrason. Ferroelectr. Freq. Control **58**(9), 1737–1750 (2011)
2. Smart Material Corp. www.smart-material.com
3. R.B. Williams, *Nonlinear mechanical and actuation characterization of piezoceramic fiber composites*. PhD thesis, Virginia Polytechnic Institute and State University (2004)
4. S.Q. Zhang, Y.X. Li, R. Schmidt, Modeling and simulation of macro-fiber composite layered smart structures. Compos. Struct. **126**, 89–100 (2015)
5. S.Q. Zhang, Z.X. Wang, X.S. Qin, G.Z. Zhao, R. Schmidt, Geometrically nonlinear analysis of composite laminated structures with multiple macro-fiber composite (MFC) actuators. Compos. Struct. **150**, 62–72 (2016)

Chapter 8
Conclusion and Future Work

Due to the excellent properties of smart structures, a wide applications can be found in aerospace, civil and automotive engineering. Smart structures are usually appearing with beam-, plate- and shell-shaped structures. There is a great concern about the appropriate computation approaches for smart structures undergo large displacement and under strong electric filed. This report dealt with geometrically nonlinear and electroelastic materially nonlinear modeling techniques for both composite laminated and piezoelectric integrated thin-walled structures.

The report first reviewed literatures on modeling and simulation techniques for smart structures. The literature survey includes through thickness hypotheses for beam, plate and shell structures, geometrically nonlinear analysis of composite and piezoelectric integrated structures, electroelastic materially nonlinear modeling methods, multi-physics coupled modeling for recently developed smart structures, and vibration control of smart structures. The literature survey reveals that most publications focused on linear modeling of piezoelectric integrated smart structure, few researchers did nonlinear analysis of smart structures, especially the electro-elastic materially nonlinear. Moreover, there are less references on multi-physics coupled simulation of newly advanced smart structures.

The second part of the report presented the mathematical preliminaries, kinematics of shell structures, and various geometrically nonlinear strain-displacement relations. Based on the first-order shear deformation hypothesis, geometrically nonlinear strain field in terms of six parameters were developed for the theories of fully geometrically nonlinear with large rotations (LRT56). Additionally, nonlinear strain field in term of five parameters were developed for the simplified nonlinear theories with the assumption of moderate rotations (RVK5, MRT5 and LRT5). Among all the mentioned geometrically nonlinear theories, LRT56 and LRT5 has fully geometrically nonlinear strain-displacement relations, MRT5 has selected nonlinear terms due to in-plane displacements, while RVK5 only includes nonlinear terms result from transverse displacements.

In the next part, constitutive relations for piezoelectric and composite materials were discussed. For deep understanding of piezo effect, the fundamental equations

S.-Q. Zhang, *Nonlinear Analysis of Thin-Walled Smart Structures*, Springer Tracts
in Mechanical Engineering, https://doi.org/10.1007/978-981-15-9857-9_8

in 3D case of piezoelectricity were first presented. Later, the electro-mechanically coupled constitutive equations were develop from 3D case to 2D plate and shell structures, in which arbitrary fibrous reinforcement orientation is considered. Based on the constitutive relations of fibrous reinforced composite materials, the constitutive models were extended to macro-fiber composite materials for two typical modes, i.e. MFC-d31 and MFC-d33. In order to compute piezoelectric materials under strong electric field, electroelastic coupled materially nonlinear constitutive equations were developed by remaining second-order terms in Taylor's expansion.

In Chap. 5, geometrically nonlinear finite element models were developed for piezoelectric bonded smart structures. Resultant stresses and strains were introduced to reduce volume integration to surface integration. The geometrically nonlinear model LRT56 not only has fully nonlinear strain-displacement relations, but also considers large rotation of shell director, in which large rotation is expressed by Euler angles. Furthermore, an eight-node shell element with 5 mechanical DOFs for each node and 1 electric DOF for each element were developed. Using the Hamilton's principle and the principle of virtual work, the dynamic and static equilibrium equations were obtained. For nonlinear case, the Total Lagrangian formulations were implemented for both geometrically and electroelastic materially nonlinear models. In the final sections, various numerical algorithms including Newmark method, central difference algorithm, Newton-Raphson method and Riks-Wempner method were developed.

The last part of the main chapters presented linear and nonlinear simulations for piezoelectric and macro-fiber composite bonded smart structures. Two major parts were investigated in this report, namely nonlinear analysis for piezoelectric integrated smart structures and nonlinear analysis for macro-fiber composite laminated structures. To validate the nonlinear FE models, geometrically nonlinear simulation of composite laminated structures were studied, including buckling analysis. Then the geometrically nonlinear FE models were applied to compute piezoelectric laminated plates and shells, as well as the electroelastic materially nonlinear models. The second part of simulation presented geometrically nonlinear analysis of macro-fiber composite integrated structures.

From the results presented in Chaps. 6 and 7, it can be concluded that simplified nonlinear shell theories, RVK5, MRT5, LRT5, will fail to predict both static and dynamic response for composite and piezoelectric laminated thin-walled structures in the range of large rotations. This is because only simplified nonlinear strain-displacement relations are considered in the RVK5 and MRT5 theories, and in addition, no proper rotation updating is possible in all these simplified nonlinear shell theories. In the case of smart structures undergoing large deflections and rotations, large rotation theory (LRT56) has to be considered. For these structures under strong driving electric field, electroelastic nonlinear phenomenon influence much on structural response, which should be considered in the simulations.

8.1 Future Research

This report developed nonlinear theories for modeling of piezoelectric integrated thin-walled smart structures. Geometrically nonlinear models include von Kármán type nonlinear, moderate rotation nonlinear, fully geometrically nonlinear with moderate rotations and large rotations. Materially nonlinear considers electroelastic nonlinearity, which permits piezo materials being under strong electric field. Even though the most important modeling techniques have been discussed in this report, there are some new theories have to develop for newly advanced smart structures, e.g. magneto-electro-elastic structures, viscoelastic smart structures. The future research interests in this field may include:

1. With the development of new materials and structures, these structures integrated not only with electric and elastic fields, but also with other physical fields, e.g. thermal, magnetic, fluid fields. Multi-physics coupled modeling techniques are quit needed for numerical simulations of the advanced smart structures.
2. Geometrically and electroelastic materially nonlinear has already considered in this report. However, for materially nonlinear effects, only electroelastic coupled nonlinear is included, many other material nonlinearities should be included in the model, for example hyperelastic nonlinear, electro-magneto-thermo-elastic coupled nonlinear, etc.
3. Nonlinearities and multi-physics coupled simulations leads to high computational consumption. High efficiency plate/shell elements based on various through thickness hypotheses should be developed.
4. New plate/shell theories should be developed and proposed for structures with large differences on each lamina, like viscoelastic piezo laminated structure.

Appendix A
Geometric Quantities

A.1 Plate Structure

The Cartesian coordinate system (X^1, X^2, X^3) and the curvilinear coordinate system $(\Theta^1, \Theta^2, \Theta^3)$ of a plate structure are shown in Fig. A.1.

The curvilinear coordinates are defined as

$$\Theta^1 = X^1, \quad \Theta^2 = X^2, \quad \Theta^3 = X^3. \tag{A.1}$$

The position vectors of an arbitrary point in the shell space and at the mid-surface are respectively expressed as

$$\boldsymbol{R} = \begin{Bmatrix} \Theta^1 \\ \Theta^2 \\ \Theta^3 \end{Bmatrix}, \quad \boldsymbol{r} = \begin{Bmatrix} \Theta^1 \\ \Theta^2 \\ 0 \end{Bmatrix}. \tag{A.2}$$

The covariant base vectors for an arbitrary point in the shell space are

$$\boldsymbol{g}_1 = \begin{Bmatrix} 1 \\ 0 \\ 0 \end{Bmatrix}, \quad \boldsymbol{g}_2 = \begin{Bmatrix} 0 \\ 1 \\ 0 \end{Bmatrix}, \quad \boldsymbol{g}_3 = \begin{Bmatrix} 0 \\ 0 \\ 1 \end{Bmatrix}. \tag{A.3}$$

The covariant and contravariant metric tensors in the shell space are

$$g_{ij} = \boldsymbol{g}_i \cdot \boldsymbol{g}_j = \begin{bmatrix} 1 & 0 & 0 \\ 0 & 1 & 0 \\ 0 & 0 & 1 \end{bmatrix}, \quad g^{ij} = [g_{ij}]^{-1} = \begin{bmatrix} 1 & 0 & 0 \\ 0 & 1 & 0 \\ 0 & 0 & 1 \end{bmatrix}. \tag{A.4}$$

S.-Q. Zhang, *Nonlinear Analysis of Thin-Walled Smart Structures*, Springer Tracts in Mechanical Engineering, https://doi.org/10.1007/978-981-15-9857-9

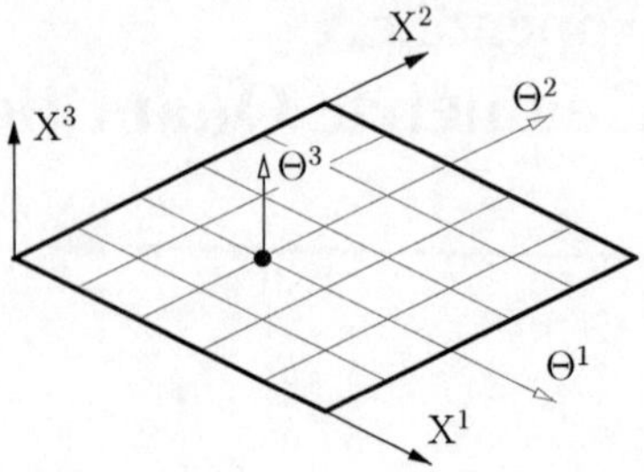

Fig. A.1 Curvilinear coordinates for a plate structure

Using the formulation $\boldsymbol{g}^i = g_{ij}\boldsymbol{g}_j$ one obtains the contravariant base vectors in the shell space

$$\boldsymbol{g}^1 = \begin{Bmatrix} 1 \\ 0 \\ 0 \end{Bmatrix}, \quad \boldsymbol{g}^2 = \begin{Bmatrix} 0 \\ 1 \\ 0 \end{Bmatrix}, \quad \boldsymbol{g}^3 = \begin{Bmatrix} 0 \\ 0 \\ 1 \end{Bmatrix}. \tag{A.5}$$

The covariant base vectors of the point at the mid-surface are

$$\boldsymbol{a}_1 = \begin{Bmatrix} 1 \\ 0 \\ 0 \end{Bmatrix}, \quad \boldsymbol{a}_2 = \begin{Bmatrix} 0 \\ 1 \\ 0 \end{Bmatrix}, \quad \boldsymbol{a}_3 = \boldsymbol{n} = \frac{\boldsymbol{a}_1 \times \boldsymbol{a}_2}{\|\boldsymbol{a}_1 \times \boldsymbol{a}_2\|} = \begin{Bmatrix} 0 \\ 0 \\ 1 \end{Bmatrix}. \tag{A.6}$$

The covariant and contravariant metric tensors at the mid-surface will be

$$a_{ij} = \boldsymbol{a}_i \cdot \boldsymbol{a}_j = \begin{bmatrix} 1\,0\,0 \\ 0\,1\,0 \\ 0\,0\,1 \end{bmatrix}, \quad a^{ij} = [a_{ij}]^{-1} = \begin{bmatrix} 1\,0\,0 \\ 0\,1\,0 \\ 0\,0\,1 \end{bmatrix}. \tag{A.7}$$

The contravariant base vectors at the mid-surface are

$$\boldsymbol{a}^1 = \begin{Bmatrix} 1 \\ 0 \\ 0 \end{Bmatrix}, \quad \boldsymbol{a}^2 = \begin{Bmatrix} 0 \\ 1 \\ 0 \end{Bmatrix}, \quad \boldsymbol{a}^3 = \boldsymbol{a}_3 = \begin{Bmatrix} 0 \\ 0 \\ 1 \end{Bmatrix}. \tag{A.8}$$

The partial derivatives of the covariant base vectors at the mid-surface are

$$\boldsymbol{a}_{i,j} = \begin{Bmatrix} 0 \\ 0 \\ 0 \end{Bmatrix}. \tag{A.9}$$

The covariant and mixed components of the curvature tensor are

$$b_{\alpha\beta} = \boldsymbol{a}_{\alpha,\beta} \cdot \boldsymbol{a}_3 = \begin{bmatrix} 0\,0 \\ 0\,0 \end{bmatrix}, \quad b_\alpha^\beta = a^{\beta\gamma} \cdot b_{\alpha\gamma} = \begin{bmatrix} 0\,0 \\ 0\,0 \end{bmatrix}. \tag{A.10}$$

The components of the shifter tensor are

$$\mu_{\alpha}^{\beta} = \delta_{\alpha}^{\beta} - \Theta^3 b_{\alpha}^{\beta} = \begin{bmatrix} 1 & 0 \\ 0 & 1 \end{bmatrix}. \tag{A.11}$$

The Christoffel symbols of the second kind for the point at the mid-surface are

$$\Gamma_{\alpha\beta}^{1} = \Gamma_{\alpha\beta}^{2} = \begin{bmatrix} 0 & 0 \\ 0 & 0 \end{bmatrix}. \tag{A.12}$$

Therefore, the covariant derivatives and the abbreviations $\overset{n}{\varphi}_{\alpha\beta}$ can be obtained as

$$\overset{n}{v}_{\alpha|\beta} = \overset{n}{v}_{\alpha,\beta}, \quad \overset{n}{\varphi}_{\alpha\beta} = \overset{n}{v}_{\alpha,\beta}, \quad \overset{n}{\varphi}_{3\alpha} = \overset{n}{v}_{3,\alpha}. \tag{A.13}$$

A.2 Cylindrical Structure

The Cartesian coordinate system (X^1, X^2, X^3) and the curvilinear coordinate system $(\Theta^1, \Theta^2, \Theta^3)$ of a cylindrical structure are shown in Fig. A.2.

The curvilinear coordinates are defined as

$$\Theta^1 = -z, \quad \Theta^2 = \alpha, \quad \Theta^3 = r - R. \tag{A.14}$$

Here, R denotes the radius of the mid-surface, and r represents the radius of an arbitrary large cylindrical surface. The position vectors of an arbitrary point in the shell space and at the mid-surface are respectively expressed as

$$\boldsymbol{R} = \begin{Bmatrix} (R + \Theta^3)\cos\left(\Theta^2\right) \\ (R + \Theta^3)\sin\left(\Theta^2\right) \\ -\Theta^1 \end{Bmatrix}, \quad \boldsymbol{r} = \begin{Bmatrix} R\cos\left(\Theta^2\right) \\ R\cos\left(\Theta^2\right) \\ -\Theta^1 \end{Bmatrix}. \tag{A.15}$$

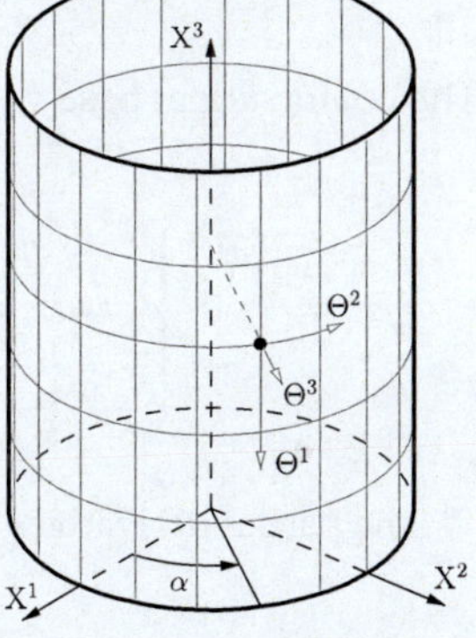

Fig. A.2 Curvilinear coordinates for a cylindrical structure

The covariant base vectors for an arbitrary point in the shell space are

$$\boldsymbol{g}_1 = \begin{Bmatrix} 0 \\ 0 \\ -1 \end{Bmatrix}, \quad \boldsymbol{g}_2 = \begin{Bmatrix} -(R+\Theta^3)\sin(\Theta^2) \\ (R+\Theta^3)\cos(\Theta^2) \\ 0 \end{Bmatrix}, \quad \boldsymbol{g}_3 = \begin{Bmatrix} \cos(\Theta^2) \\ \sin(\Theta^2) \\ 0 \end{Bmatrix}. \tag{A.16}$$

The covariant and contravariant metric tensors in the shell space are

$$g_{ij} = \boldsymbol{g}_i \cdot \boldsymbol{g}_j = \begin{bmatrix} 1 & 0 & 0 \\ 0 & (R+\Theta^3)^2 & 0 \\ 0 & 0 & 1 \end{bmatrix}, \quad g^{ij} = [g_{ij}]^{-1} = \begin{bmatrix} 1 & 0 & 0 \\ 0 & \dfrac{1}{(R+\Theta^3)^2} & 0 \\ 0 & 0 & 1 \end{bmatrix}. \tag{A.17}$$

The contravariant base vectors in the shell space are

$$\boldsymbol{g}^1 = \begin{Bmatrix} 0 \\ 0 \\ -1 \end{Bmatrix}, \quad \boldsymbol{g}^2 = \begin{Bmatrix} -\dfrac{\sin(\Theta^2)}{(R+\Theta^3)} \\ \dfrac{\cos(\Theta^2)}{(R+\Theta^3)} \\ 0 \end{Bmatrix}, \quad \boldsymbol{g}^3 = \begin{Bmatrix} \cos(\Theta^2) \\ \sin(\Theta^2) \\ 0 \end{Bmatrix}. \tag{A.18}$$

The covariant base vectors at the mid-surface are

$$\boldsymbol{a}_1 = \begin{Bmatrix} 0 \\ 0 \\ -1 \end{Bmatrix}, \quad \boldsymbol{a}_2 = \begin{Bmatrix} -R\sin(\Theta^2) \\ R\cos(\Theta^2) \\ 0 \end{Bmatrix}, \quad \boldsymbol{a}_3 = \boldsymbol{n} = \begin{Bmatrix} \cos(\Theta^2) \\ \sin(\Theta^2) \\ 0 \end{Bmatrix}. \tag{A.19}$$

The covariant and contravariant metric tensors at the mid-surface are

$$a_{ij} = \boldsymbol{a}_i \cdot \boldsymbol{a}_j = \begin{bmatrix} 1 & 0 & 0 \\ 0 & R^2 & 0 \\ 0 & 0 & 1 \end{bmatrix}, \quad a^{ij} = [a_{ij}]^{-1} = \begin{bmatrix} 1 & 0 & 0 \\ 0 & \dfrac{1}{R^2} & 0 \\ 0 & 0 & 1 \end{bmatrix}. \tag{A.20}$$

The contravariant base vectors at the mid-surface are

$$\boldsymbol{a}^1 = \begin{Bmatrix} 0 \\ 0 \\ -1 \end{Bmatrix}, \quad \boldsymbol{a}^2 = \begin{Bmatrix} -\dfrac{\sin(\Theta^2)}{R} \\ \dfrac{\cos(\Theta^2)}{R} \\ 0 \end{Bmatrix}, \quad \boldsymbol{a}^3 = \boldsymbol{a}_3 = \begin{Bmatrix} \cos(\Theta^2) \\ \sin(\Theta^2) \\ 0 \end{Bmatrix}. \tag{A.21}$$

The partial derivatives of the covariant base vectors at the mid-surface are

$$\boldsymbol{a}_{1,1} = \boldsymbol{a}_{1,2} = \boldsymbol{a}_{1,3} = \boldsymbol{a}_{2,3} = \begin{Bmatrix} 0 \\ 0 \\ 0 \end{Bmatrix}, \ \boldsymbol{a}_{2,2} = \begin{Bmatrix} -R\cos\left(\Theta^2\right) \\ -R\sin\left(\Theta^2\right) \\ 0 \end{Bmatrix}, \ \boldsymbol{a}_{3,2} = \begin{Bmatrix} -\sin\left(\Theta^2\right) \\ \cos\left(\Theta^2\right) \\ 0 \end{Bmatrix}. \tag{A.22}$$

The covariant and mixed components of the curvature tensor are

$$b_{\alpha\beta} = \boldsymbol{a}_{\alpha,\beta} \cdot \boldsymbol{a}_3 = \begin{bmatrix} 0 & 0 \\ 0 & -R \end{bmatrix}, \quad b_\alpha^\beta = a^{\beta\gamma} \cdot b_{\alpha\gamma} = \begin{bmatrix} 0 & 0 \\ 0 & -\frac{1}{R} \end{bmatrix}. \tag{A.23}$$

The components of the shifter tensor are

$$\mu_\alpha^\beta = \delta_\alpha^\beta - \Theta^3 b_\alpha^\beta = \begin{bmatrix} 1 & 0 \\ 0 & 1 + \frac{\Theta^3}{R} \end{bmatrix}. \tag{A.24}$$

The Christoffel symbols of the second kind for the point at the mid-surface are

$$\Gamma_{\alpha\beta}^1 = \Gamma_{\alpha\beta}^2 = \begin{bmatrix} 0 & 0 \\ 0 & 0 \end{bmatrix}. \tag{A.25}$$

Therefore, the covariant derivatives and the abbreviations $\overset{n}{\varphi}_{\alpha\beta}$ can be obtained as

$$\begin{cases} \overset{n}{v}_{1|1} = \overset{n}{v}_{1,1} \\ \overset{n}{v}_{1|2} = \overset{n}{v}_{1,2} \\ \overset{n}{v}_{2|1} = \overset{n}{v}_{2,1} \\ \overset{n}{v}_{2|2} = \overset{n}{v}_{2,2} \end{cases} \text{ and } \begin{cases} \overset{n}{\varphi}_{11} = \overset{n}{v}_{1,1} \\ \overset{n}{\varphi}_{12} = \overset{n}{v}_{1,2} \\ \overset{n}{\varphi}_{21} = \overset{n}{v}_{2,1} \\ \overset{n}{\varphi}_{22} = \overset{n}{v}_{2,2} + R\overset{n}{v}_3 \\ \overset{n}{\varphi}_{31} = \overset{n}{v}_{3,1} \\ \overset{n}{\varphi}_{32} = \overset{n}{v}_{3,2} - \frac{1}{R}\overset{n}{v}_2 \end{cases} \tag{A.26}$$

A.3 Spherical Structure

The Cartesian coordinate system (X^1, X^2, X^3) and the curvilinear coordinate system $(\Theta^1, \Theta^2, \Theta^3)$ of a spherical structure are shown in Fig. A.3.

Here, R denotes the radius of the mid-surface, and r is the radius of an arbitrary large spherical surface. The curvilinear coordinates are defined as

$$\Theta^1 = R\beta, \quad \Theta^2 = \alpha, \quad \Theta^3 = r - R. \tag{A.27}$$

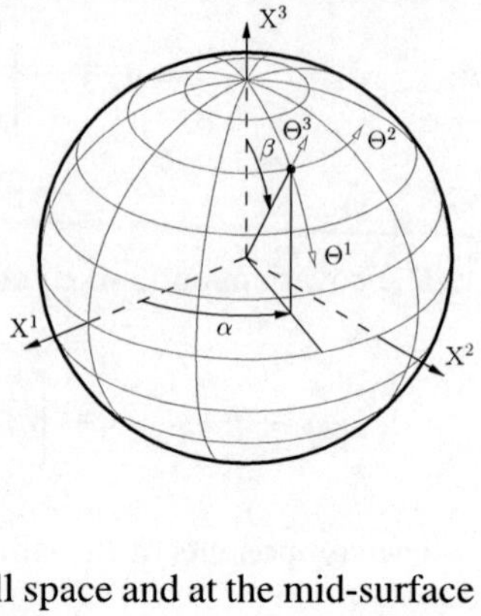

Fig. A.3 Curvilinear coordinates for a spherical structure

The position vectors of an arbitrary point in the shell space and at the mid-surface are respectively expressed as

$$\boldsymbol{R} = (R + \Theta^3)\begin{Bmatrix} \sin\left(\dfrac{\Theta^1}{R}\right)\cos\left(\Theta^2\right) \\ \sin\left(\dfrac{\Theta^1}{R}\right)\cos\left(\Theta^2\right) \\ \cos\left(\dfrac{\Theta^1}{R}\right) \end{Bmatrix}, \quad \boldsymbol{r} = R\begin{Bmatrix} \sin\left(\dfrac{\Theta^1}{R}\right)\cos\left(\Theta^2\right) \\ \sin\left(\dfrac{\Theta^1}{R}\right)\cos\left(\Theta^2\right) \\ \cos\left(\dfrac{\Theta^1}{R}\right) \end{Bmatrix}. \tag{A.28}$$

The covariant base vectors for an arbitrary point in the shell space are

$$\boldsymbol{g}_1 = \left(1 + \frac{\Theta^3}{R}\right)\begin{Bmatrix} \cos\left(\dfrac{\Theta^1}{R}\right)\cos\left(\Theta^2\right) \\ \cos\left(\dfrac{\Theta^1}{R}\right)\sin\left(\Theta^2\right) \\ -\sin\left(\dfrac{\Theta^1}{R}\right) \end{Bmatrix},$$

$$\boldsymbol{g}_2 = (R + \Theta^3)\begin{Bmatrix} -\sin\left(\dfrac{\Theta^1}{R}\right)\sin\left(\Theta^2\right) \\ \sin\left(\dfrac{\Theta^1}{R}\right)\cos\left(\Theta^2\right) \\ 0 \end{Bmatrix}, \quad \boldsymbol{g}_3 = \begin{Bmatrix} \sin\left(\dfrac{\Theta^1}{R}\right)\cos\left(\Theta^2\right) \\ \sin\left(\dfrac{\Theta^1}{R}\right)\sin\left(\Theta^2\right) \\ \cos\left(\dfrac{\Theta^1}{R}\right) \end{Bmatrix}. \tag{A.29}$$

The covariant metric tensor in the shell space is

$$g_{ij} = \boldsymbol{g}_i \cdot \boldsymbol{g}_j = \begin{bmatrix} \left(1 + \dfrac{\Theta^3}{R}\right)^2 & 0 & 0 \\ 0 & (R + \Theta^3)^2 \sin^2\left(\dfrac{\Theta^1}{R}\right) & 0 \\ 0 & 0 & 1 \end{bmatrix}. \tag{A.30}$$

The contravariant metric tensor in the shell space is

$$g^{ij} = [g_{ij}]^{-1} = \begin{bmatrix} \dfrac{R^2}{\left(R+\Theta^3\right)^2} & 0 & 0 \\ 0 & \dfrac{1}{\left(R+\Theta^3\right)^2 \sin^2\left(\dfrac{\Theta^1}{R}\right)} & 0 \\ 0 & 0 & 1 \end{bmatrix}. \tag{A.31}$$

The contravariant base vectors in the shell space are

$$\boldsymbol{g}^1 = \frac{R}{R+\Theta^3} \begin{Bmatrix} \cos\left(\dfrac{\Theta^1}{R}\right)\cos\left(\Theta^2\right) \\ \cos\left(\dfrac{\Theta^1}{R}\right)\sin\left(\Theta^2\right) \\ -\sin\left(\dfrac{\Theta^1}{R}\right) \end{Bmatrix},$$

$$\boldsymbol{g}^2 = \frac{1}{R+\Theta^3} \begin{Bmatrix} -\dfrac{\sin\left(\Theta^2\right)}{\sin\left(\dfrac{\Theta^1}{R}\right)} \\ \dfrac{\cos\left(\Theta^2\right)}{\sin\left(\dfrac{\Theta^1}{R}\right)} \\ 0 \end{Bmatrix}, \quad \boldsymbol{g}^3 = \begin{Bmatrix} \sin\left(\dfrac{\Theta^1}{R}\right)\cos\left(\Theta^2\right) \\ \sin\left(\dfrac{\Theta^1}{R}\right)\sin\left(\Theta^2\right) \\ \cos\left(\dfrac{\Theta^1}{R}\right) \end{Bmatrix}. \tag{A.32}$$

The covariant base vectors at the mid-surface are

$$\boldsymbol{a}_1 = \begin{Bmatrix} \cos\left(\dfrac{\Theta^1}{R}\right)\cos\left(\Theta^2\right) \\ \cos\left(\dfrac{\Theta^1}{R}\right)\sin\left(\Theta^2\right) \\ -\sin\left(\dfrac{\Theta^1}{R}\right) \end{Bmatrix}, \quad \boldsymbol{a}_2 = R \begin{Bmatrix} -\sin\left(\dfrac{\Theta^1}{R}\right)\sin\left(\Theta^2\right) \\ \sin\left(\dfrac{\Theta^1}{R}\right)\cos\left(\Theta^2\right) \\ 0 \end{Bmatrix},$$

$$\boldsymbol{a}_3 = \boldsymbol{n} = \begin{Bmatrix} \sin\left(\dfrac{\Theta^1}{R}\right)\cos\left(\Theta^2\right) \\ \sin\left(\dfrac{\Theta^1}{R}\right)\sin\left(\Theta^2\right) \\ \cos\left(\dfrac{\Theta^1}{R}\right) \end{Bmatrix}. \tag{A.33}$$

The covariant and contravariant metric tensors at the mid-surface are

$$a_{ij} = \boldsymbol{a}_i \cdot \boldsymbol{a}_j = \begin{bmatrix} 1 & 0 & 0 \\ 0 & R^2 \sin^2\left(\frac{\Theta^1}{R}\right) & 0 \\ 0 & 0 & 1 \end{bmatrix}, \quad a^{ij} = a_{ij}^{-1} = \begin{bmatrix} 1 & 0 & 0 \\ 0 & \dfrac{1}{R^2 \sin^2\left(\frac{\Theta^1}{R}\right)} & 0 \\ 0 & 0 & 1 \end{bmatrix}. \tag{A.34}$$

The contravariant base vectors at the mid-surface are

$$\boldsymbol{a}^1 = \begin{Bmatrix} \cos\left(\frac{\Theta^1}{R}\right)\cos\left(\Theta^2\right) \\ \cos\left(\frac{\Theta^1}{R}\right)\sin\left(\Theta^2\right) \\ -\sin\left(\frac{\Theta^1}{R}\right) \end{Bmatrix}, \quad \boldsymbol{a}^2 = \begin{Bmatrix} -\dfrac{\sin\left(\Theta^2\right)}{\sin\left(\frac{\Theta^1}{R}\right)} \\ \dfrac{\cos\left(\Theta^2\right)}{\sin\left(\frac{\Theta^1}{R}\right)} \\ 0 \end{Bmatrix}, \quad \boldsymbol{a}^3 = \begin{Bmatrix} \sin\left(\frac{\Theta^1}{R}\right)\cos\left(\Theta^2\right) \\ \sin\left(\frac{\Theta^1}{R}\right)\sin\left(\Theta^2\right) \\ \cos\left(\frac{\Theta^1}{R}\right) \end{Bmatrix}. \tag{A.35}$$

The partial derivatives of the covariant base vectors at the mid-surface are

$$\boldsymbol{a}_{1,1} = -\frac{1}{R}\begin{Bmatrix} \sin\left(\frac{\Theta^1}{R}\right)\cos\left(\Theta^2\right) \\ \sin\left(\frac{\Theta^1}{R}\right)\sin\left(\Theta^2\right) \\ \cos\left(\frac{\Theta^1}{R}\right) \end{Bmatrix}, \quad \boldsymbol{a}_{1,2} = \boldsymbol{a}_{2,1} = \begin{Bmatrix} -\cos\left(\frac{\Theta^1}{R}\right)\sin\left(\Theta^2\right) \\ \cos\left(\frac{\Theta^1}{R}\right)\cos\left(\Theta^2\right) \\ 0 \end{Bmatrix},$$

$$\boldsymbol{a}_{2,2} = \begin{Bmatrix} -\sin\left(\frac{\Theta^1}{R}\right)\cos\left(\Theta^2\right) \\ -\sin\left(\frac{\Theta^1}{R}\right)\sin\left(\Theta^2\right) \\ 0 \end{Bmatrix}, \quad \boldsymbol{a}_{3,1} = \frac{1}{R}\begin{Bmatrix} \cos\left(\frac{\Theta^1}{R}\right)\cos\left(\Theta^2\right) \\ \cos\left(\frac{\Theta^1}{R}\right)\sin\left(\Theta^2\right) \\ -\sin\left(\frac{\Theta^1}{R}\right) \end{Bmatrix},$$

$$\boldsymbol{a}_{3,2} = \begin{Bmatrix} -\sin\left(\frac{\Theta^1}{R}\right)\sin\left(\Theta^2\right) \\ \sin\left(\frac{\Theta^1}{R}\right)\cos\left(\Theta^2\right) \\ 0 \end{Bmatrix}, \quad \boldsymbol{a}_{1,3} = \boldsymbol{a}_{2,3} = \boldsymbol{a}_{3,3} = \begin{Bmatrix} 0 \\ 0 \\ 0 \end{Bmatrix}. \tag{A.36}$$

The covariant and mixed components of the curvature tensor are

$$b_{\alpha\beta} = \begin{bmatrix} -\frac{1}{R} & 0 \\ 0 & -R\sin^2\left(\frac{\Theta^1}{R}\right) \end{bmatrix}, \quad b_\alpha^\beta = \begin{bmatrix} -\frac{1}{R} & 0 \\ 0 & -\frac{1}{R} \end{bmatrix}. \tag{A.37}$$

The components of the shifter tensor are

$$\mu_\alpha^\beta = \delta_\alpha^\beta - \Theta^3 \cdot b_\alpha^\beta = \begin{bmatrix} 1 + \frac{\Theta^3}{R} & 0 \\ 0 & 1 + \frac{\Theta^3}{R} \end{bmatrix}. \tag{A.38}$$

The Christoffel symbols of the second kind for the point at the mid-surface are

$$\Gamma_{\alpha\beta}^1 = \begin{bmatrix} 0 & 0 \\ 0 & -R\sin\left(\frac{\Theta^1}{R}\right)\cos\left(\frac{\Theta^1}{R}\right) \end{bmatrix}, \quad \Gamma_{\alpha\beta}^2 = \begin{bmatrix} 0 & \dfrac{\cos\left(\frac{\Theta^1}{R}\right)}{R\sin\left(\frac{\Theta^1}{R}\right)} \\ \dfrac{\cos\left(\frac{\Theta^1}{R}\right)}{R\sin\left(\frac{\Theta^1}{R}\right)} & 0 \end{bmatrix}. \tag{A.39}$$

Therefore, the covariant derivatives and the abbreviations $\overset{n}{\varphi}_{\alpha\beta}$ can be obtained as

$$\begin{cases} \overset{n}{v}_{1|1} = \overset{n}{v}_{1,1} \\ \overset{n}{v}_{1|2} = \overset{n}{v}_{1,2} - \dfrac{\cos\left(\frac{\Theta^1}{R}\right)}{R\sin\left(\frac{\Theta^1}{R}\right)} \overset{n}{v}_2 \\ \overset{n}{v}_{2|1} = \overset{n}{v}_{2,1} - \dfrac{\cos\left(\frac{\Theta^1}{R}\right)}{R\sin\left(\frac{\Theta^1}{R}\right)} \overset{n}{v}_2 \\ \overset{n}{v}_{2|2} = \overset{n}{v}_{2,2} + R\sin\left(\frac{\Theta^1}{R}\right)\cos\left(\frac{\Theta^1}{R}\right) \overset{n}{v}_1 \end{cases} \text{and} \begin{cases} \overset{n}{\varphi}_{11} = \overset{n}{v}_{1,1} + \frac{1}{R} \overset{n}{v}_3 \\ \overset{n}{\varphi}_{12} = \overset{n}{v}_{1,2} - \dfrac{\cos\left(\frac{\Theta^1}{R}\right)}{R\sin\left(\frac{\Theta^1}{R}\right)} \overset{n}{v}_2 \\ \overset{n}{\varphi}_{21} = \overset{n}{v}_{2,1} - \dfrac{\cos\left(\frac{\Theta^1}{R}\right)}{R\sin\left(\frac{\Theta^1}{R}\right)} \overset{n}{v}_2 \\ \overset{n}{\varphi}_{22} = \overset{n}{v}_{2,2} + R\sin\left(\frac{\Theta^1}{R}\right)\cos\left(\frac{\Theta^1}{R}\right) \overset{n}{v}_1 \\ \qquad + R\sin^2\left(\frac{\Theta^1}{R}\right) \overset{n}{v}_3 \\ \overset{n}{\varphi}_{31} = \overset{n}{v}_{3,1} - \frac{1}{R}\overset{n}{v}_1 \\ \overset{n}{\varphi}_{32} = \overset{n}{v}_{3,2} - \frac{1}{R}\overset{n}{v}_2 \end{cases} \tag{A.40}$$

We introduce several variables that are frequently used in the strain-displacement expressions, as shown in Table A.1.

Table A.1 Notations of frequently used geometric quantities

Notation	Sphere	Cylinder	Plate
$a_1 = a^{11}$	1	1	1
$a_2 = a^{22}$	$\dfrac{1}{R^2 \sin^2\left(\frac{\Theta^1}{R}\right)}$	$\dfrac{1}{R^2}$	1
$b_1 = b_{11}$	$-\dfrac{1}{R}$	0	0
$b_2 = b_{22}$	$-R\sin^2\left(\dfrac{\Theta^1}{R}\right)$	$-R$	0
$t_1 = -\Gamma^2_{12} = -\Gamma^2_{21}$	$-\dfrac{\cos\left(\frac{\Theta^1}{R}\right)}{R\sin\left(\frac{\Theta^1}{R}\right)}$	0	0
$t_2 = -\Gamma^1_{22}$	$R\sin\left(\frac{\Theta^1}{R}\right)\cos\left(\frac{\Theta^1}{R}\right)$	0	0
$c_1 = b^1_{.1}$	$c_1 = a_1 b_1$		
$c_2 = b^2_{.2}$	$c_2 = a_2 b_2$		

Using the notations introduced in Table A.1, a general expression of $\overset{n}{\varphi}_{\alpha\beta}$ for the three curvilinear coordinate systems can be obtained as

$$\begin{cases} \overset{n}{\varphi}_{11} = & \overset{n}{v}_{1,1} - b_{11}\overset{n}{v}_3 & = \overset{n}{v}_{1,1} - b_1\overset{n}{v}_3 \\ \overset{n}{\varphi}_{12} = & \overset{n}{v}_{1,2} - \Gamma^2_{12}\overset{n}{v}_2 & = \overset{n}{v}_{1,2} + t_1\overset{n}{v}_2 \\ \overset{n}{\varphi}_{21} = & \overset{n}{v}_{2,1} - \Gamma^2_{21}\overset{n}{v}_2 & = \overset{n}{v}_{2,1} + t_1\overset{n}{v}_2 \\ \overset{n}{\varphi}_{22} = & \overset{n}{v}_{2,2} - \Gamma^1_{22}\overset{n}{v}_1 - b_{22}\overset{n}{v}_3 & = \overset{n}{v}_{2,2} + t_2\overset{n}{v}_1 - b_2\overset{n}{v}_3 \\ \overset{n}{\varphi}_{31} = & \overset{n}{v}_{3,1} + b^1_{.1}\overset{n}{v}_1 & = \overset{n}{v}_{3,1} + c_1\overset{n}{v}_1 \\ \overset{n}{\varphi}_{32} = & \overset{n}{v}_{3,2} + b^2_{.2}\overset{n}{v}_2 & = \overset{n}{v}_{3,2} + c_2\overset{n}{v}_2 \end{cases} \tag{A.41}$$

Appendix B
Strain Fields of LRT56 Theory

The Green-Lagrange strain terms ($\overset{n}{\varepsilon}_{\alpha\beta}$, $\overset{n}{\varepsilon}_{\alpha 3}$ and $\overset{0}{\varepsilon}_{33} = 0$) can be derived as

$$\begin{aligned}
\overset{0}{\varepsilon}_{11} &= \overset{0}{v}_{1,1} - b_1\overset{0}{v}_3 \quad \text{(LIN5)} \\
&+ \frac{1}{2}\left[a_1(\overset{0}{v}_{1,1})^2 + a_1(b_1)^2(\overset{0}{v}_3)^2 + a_2(\overset{0}{v}_{2,1})^2 + a_2(t_1)^2(\overset{0}{v}_2)^2 + (\overset{0}{v}_{3,1})^2 + (c_1)^2(\overset{0}{v}_1)^2\right] \\
&+ \frac{1}{2}\left[-a_1b_1\overset{0}{v}_{1,1}\overset{0}{v}_3 - a_1b_1\overset{0}{v}_3\overset{0}{v}_{1,1} + a_2t_1\overset{0}{v}_{2,1}\overset{0}{v}_2 + a_2t_1\overset{0}{v}_2\overset{0}{v}_{2,1} + c_1\overset{0}{v}_{3,1}\overset{0}{v}_1 + c_1\overset{0}{v}_1\overset{0}{v}_{3,1}\right]
\end{aligned} \tag{B.1}$$

$$\begin{aligned}
\overset{0}{\varepsilon}_{22} &= \overset{0}{v}_{2,2} + t_2\overset{0}{v}_1 - b_2\overset{0}{v}_3 \quad \text{(LIN5)} \\
&+ \frac{1}{2}\Big[a_1(\overset{0}{v}_{1,2})^2 + \left(a_1(t_1)^2 + (c_2)^2\right)(\overset{0}{v}_2)^2 + a_2(\overset{0}{v}_{2,2})^2 + a_2(t_2)^2(\overset{0}{v}_1)^2 \\
&+ a_2(b_2)^2(\overset{0}{v}_3)^2 + (\overset{0}{v}_{3,2})^2\Big] \\
&+ \frac{1}{2}\left[a_1t_1\overset{0}{v}_{1,2}\overset{0}{v}_2 + a_1t_1\overset{0}{v}_2\overset{0}{v}_{1,2} + a_2t_2\overset{0}{v}_{2,2}\overset{0}{v}_1 + a_2t_2\overset{0}{v}_1\overset{0}{v}_{2,2}\right] \\
&+ \frac{1}{2}\left[-a_2b_2\overset{0}{v}_{2,2}\overset{0}{v}_3 - a_2b_2\overset{0}{v}_3\overset{0}{v}_{2,2} + c_2\overset{0}{v}_{3,2}\overset{0}{v}_2 + c_2\overset{0}{v}_2\overset{0}{v}_{3,2}\right] \\
&+ \frac{1}{2}\left[-a_2t_2b_2\overset{0}{v}_1\overset{0}{v}_3 - a_2t_2b_2\overset{0}{v}_3\overset{0}{v}_1\right]
\end{aligned} \tag{B.2}$$

S.-Q. Zhang, *Nonlinear Analysis of Thin-Walled Smart Structures*, Springer Tracts
in Mechanical Engineering, https://doi.org/10.1007/978-981-15-9857-9

$$
\begin{aligned}
2\overset{0}{\varepsilon}_{12} = {} & \overset{0}{v}_{1,2} + \overset{0}{v}_{2,1} + 2t_1\overset{0}{v}_2 \quad \text{(LIN5)} \\
& + \frac{1}{2}\Big[a_1\overset{0}{v}_{1,2}\overset{0}{v}_{1,1} + a_1\overset{0}{v}_{1,1}\overset{0}{v}_{1,2} - (a_1t_1b_1 + a_2b_2t_1)\,\overset{0}{v}_2\overset{0}{v}_3 - (a_1t_1b_1 + a_2b_2t_1)\,\overset{0}{v}_3\overset{0}{v}_2 \\
& \qquad + a_2\overset{0}{v}_{2,2}\overset{0}{v}_{2,1} + a_2\overset{0}{v}_{2,1}\overset{0}{v}_{2,2} + \overset{0}{v}_{3,2}\overset{0}{v}_{3,1} + \overset{0}{v}_{3,1}\overset{0}{v}_{3,2}\Big] \\
& + \frac{1}{2}\Big[-a_1b_1\overset{0}{v}_{1,2}\overset{0}{v}_3 - a_1b_1\overset{0}{v}_3\overset{0}{v}_{1,2} + a_1t_1\overset{0}{v}_2\overset{0}{v}_{1,1} + a_1t_1\overset{0}{v}_{1,1}\overset{0}{v}_2 + a_2t_2\overset{0}{v}_1\overset{0}{v}_{2,1} + a_2t_2\overset{0}{v}_{2,1}\overset{0}{v}_1\Big] \\
& + \frac{1}{2}\Big[a_2t_1\overset{0}{v}_{2,2}\overset{0}{v}_2 + a_2t_1\overset{0}{v}_2\overset{0}{v}_{2,2} - a_2b_2\overset{0}{v}_3\overset{0}{v}_{2,1} - a_2b_2\overset{0}{v}_{2,1}\overset{0}{v}_3 + c_1\overset{0}{v}_{3,2}\overset{0}{v}_1 + c_1\overset{0}{v}_1\overset{0}{v}_{3,2}\Big] \\
& + \frac{1}{2}\Big[(a_2t_2t_1 + c_1c_2)\,\overset{0}{v}_1\overset{0}{v}_2 + (a_2t_2t_1 + c_1c_2)\,\overset{0}{v}_2\overset{0}{v}_1\Big] \\
& + \frac{1}{2}\Big[c_2\overset{0}{v}_2\overset{0}{v}_{3,1} + c_2\overset{0}{v}_{3,1}\overset{0}{v}_2\Big]
\end{aligned}
\tag{B.3}
$$

$$
\begin{aligned}
\overset{1}{\varepsilon}_{11} = {} & \overset{1}{v}_{1,1} - a_1b_1\overset{0}{v}_{1,1} + a_1(b_1)^2\overset{0}{v}_3 - b_1\overset{1}{v}_3 \quad \text{(LIN5)} \\
& + \frac{1}{2}\Big[a_1\overset{0}{v}_{1,1}\overset{1}{v}_{1,1} + a_1\overset{1}{v}_{1,1}\overset{0}{v}_{1,1} + a_2\overset{0}{v}_{2,1}\overset{1}{v}_{2,1} + a_2\overset{1}{v}_{2,1}\overset{0}{v}_{2,1} + a_2(t_1)^2\overset{0}{v}_2\overset{1}{v}_2 + a_2(t_1)^2\overset{1}{v}_2\overset{0}{v}_2 \\
& \qquad + c_1\overset{0}{v}_{3,1}\overset{1}{v}_1 + c_1\overset{1}{v}_1\overset{0}{v}_{3,1} + a_1(b_1)^2\overset{0}{v}_3\overset{1}{v}_3 + a_1(b_1)^2\overset{1}{v}_3\overset{0}{v}_3 + c_1\overset{0}{v}_1\overset{1}{v}_{3,1} + c_1\overset{1}{v}_{3,1}\overset{0}{v}_1\Big] \\
& + \frac{1}{2}\Big[-a_1b_1\overset{0}{v}_3\overset{1}{v}_{1,1} - a_1b_1\overset{1}{v}_{1,1}\overset{0}{v}_3 + a_2t_1\overset{0}{v}_{2,1}\overset{1}{v}_2 + a_2t_1\overset{1}{v}_2\overset{0}{v}_{2,1} + a_2t_1\overset{0}{v}_2\overset{1}{v}_{2,1} + a_2t_1\overset{1}{v}_{2,1}\overset{0}{v}_2 \\
& \qquad + (c_1)^2\overset{0}{v}_1\overset{1}{v}_1 + (c_1)^2\overset{1}{v}_1\overset{0}{v}_1 - a_1b_1\overset{0}{v}_{1,1}\overset{1}{v}_3 - a_1b_1\overset{1}{v}_3\overset{0}{v}_{1,1} + \overset{0}{v}_{3,1}\overset{1}{v}_{3,1} + \overset{1}{v}_{3,1}\overset{0}{v}_{3,1}\Big]
\end{aligned}
\tag{B.4}
$$

$$
\begin{aligned}
\overset{1}{\varepsilon}_{22} = {} & \overset{1}{v}_{2,2} + t_2\overset{1}{v}_1 - a_2b_2\overset{0}{v}_{2,2} - a_2b_2t_2\overset{0}{v}_1 + a_2(b_2)^2\overset{0}{v}_3 - b_2\overset{1}{v}_3 \quad \text{(LIN5)} \\
& + \frac{1}{2}\Big[a_1\overset{0}{v}_{1,2}\overset{1}{v}_{1,2} + a_1\overset{1}{v}_{1,2}\overset{0}{v}_{1,2} + \left(a_1(t_1)^2 + (c_2)^2\right)\overset{0}{v}_2\overset{1}{v}_2 + \left(a_1(t_1)^2 + (c_2)^2\right)\overset{1}{v}_2\overset{0}{v}_2 \\
& \qquad + a_2\overset{0}{v}_{2,2}\overset{1}{v}_{2,2} + a_2\overset{1}{v}_{2,2}\overset{0}{v}_{2,2} + a_2(t_2)^2\overset{0}{v}_1\overset{1}{v}_1 + a_2(t_2)^2\overset{1}{v}_1\overset{0}{v}_1 + a_2(b_2)^2\overset{0}{v}_3\overset{1}{v}_3 \\
& \qquad + a_2(b_2)^2\overset{1}{v}_3\overset{0}{v}_3 + \overset{0}{v}_{3,2}\overset{1}{v}_{3,2} + \overset{1}{v}_{3,2}\overset{0}{v}_{3,2}\Big] \\
& + \frac{1}{2}\Big[a_1t_1\overset{0}{v}_{1,2}\overset{1}{v}_2 + a_1t_1\overset{1}{v}_2\overset{0}{v}_{1,2} + a_1t_1\overset{0}{v}_2\overset{1}{v}_{1,2} + a_1t_1\overset{1}{v}_{1,2}\overset{0}{v}_2 + a_2t_2\overset{0}{v}_{2,2}\overset{1}{v}_1 + a_2t_2\overset{1}{v}_1\overset{0}{v}_{2,2} \\
& \qquad + a_2t_2\overset{0}{v}_1\overset{1}{v}_{2,2} + a_2t_2\overset{1}{v}_{2,2}\overset{0}{v}_1\Big] \\
& + \frac{1}{2}\Big[-a_2b_2\overset{0}{v}_3\overset{1}{v}_{2,2} - a_2b_2\overset{1}{v}_{2,2}\overset{0}{v}_3 + c_2\overset{0}{v}_{3,2}\overset{1}{v}_2 + c_2\overset{1}{v}_2\overset{0}{v}_{3,2} - a_2b_2\overset{0}{v}_{2,2}\overset{1}{v}_3 - a_2b_2\overset{1}{v}_3\overset{0}{v}_{2,2} \\
& \qquad + c_2\overset{0}{v}_2\overset{1}{v}_{3,2} + c_2\overset{1}{v}_{3,2}\overset{0}{v}_2\Big] \\
& + \frac{1}{2}\Big[-a_2b_2t_2\overset{0}{v}_3\overset{1}{v}_1 - a_2b_2t_2\overset{1}{v}_1\overset{0}{v}_3 - a_2b_2t_2\overset{1}{v}_3\overset{0}{v}_1 - a_2b_2t_2\overset{0}{v}_1\overset{1}{v}_3\Big]
\end{aligned}
\tag{B.5}
$$

$$
\begin{aligned}
2\overset{1}{\varepsilon}_{12} = {} & \overset{1}{v}_{1,2} + \overset{1}{v}_{2,1} + 2t_1\overset{1}{v}_2 - a_2b_2\overset{0}{v}_{2,1} - a_1b_1\overset{0}{v}_{1,2} - (a_2b_2t_1 + a_1b_1t_1)\,\overset{0}{v}_2 \quad \text{(LIN5)} \\
& + \frac{1}{2}\Big[a_1\overset{0}{v}_{1,1}\overset{1}{v}_{1,2} + a_1\overset{1}{v}_{1,2}\overset{0}{v}_{1,1} - a_1b_1t_1\overset{0}{v}_3\overset{1}{v}_2 - a_1b_1t_1\overset{1}{v}_2\overset{0}{v}_3 + a_2\overset{0}{v}_{2,1}\overset{1}{v}_{2,2} + a_2\overset{1}{v}_{2,2}\overset{0}{v}_{2,1} \\
& \qquad + (a_2t_1t_2 + c_1c_2)\,\overset{1}{v}_1\overset{0}{v}_2 + (a_2t_1t_2 + c_1c_2)\,\overset{0}{v}_2\overset{1}{v}_1 + a_1\overset{1}{v}_{1,1}\overset{0}{v}_{1,2} + a_1\overset{0}{v}_{1,2}\overset{1}{v}_{1,1} \\
& \qquad + a_2t_2\overset{1}{v}_{2,1}\overset{0}{v}_1 + a_2t_2\overset{0}{v}_1\overset{1}{v}_{2,1} + \overset{0}{v}_{3,1}\overset{1}{v}_{3,2} + \overset{1}{v}_{3,2}\overset{0}{v}_{3,1} + \overset{1}{v}_{3,1}\overset{0}{v}_{3,2} + \overset{0}{v}_{3,2}\overset{1}{v}_{3,1}\Big] \\
& + \frac{1}{2}\Big[a_1t_1\overset{0}{v}_{1,1}\overset{1}{v}_2 + a_1t_1\overset{1}{v}_2\overset{0}{v}_{1,1} - a_1b_1\overset{0}{v}_3\overset{1}{v}_{1,2} - a_1b_1\overset{1}{v}_{1,2}\overset{0}{v}_3 + a_2t_2\overset{0}{v}_{2,1}\overset{1}{v}_1 + a_2t_2\overset{1}{v}_1\overset{0}{v}_{2,1} \\
& \qquad + a_2t_1\overset{0}{v}_2\overset{1}{v}_{2,2} + a_2t_1\overset{1}{v}_{2,2}\overset{0}{v}_2 + c_1\overset{0}{v}_1\overset{1}{v}_{3,2} + c_1\overset{1}{v}_{3,2}\overset{0}{v}_1 - a_1b_1\overset{1}{v}_3\overset{0}{v}_{1,2} - a_1b_1\overset{0}{v}_{1,2}\overset{1}{v}_3\Big] \\
& + \frac{1}{2}\Big[c_2\overset{0}{v}_{3,1}\overset{1}{v}_2 + c_2\overset{1}{v}_2\overset{0}{v}_{3,1} + a_1t_1\overset{1}{v}_{1,1}\overset{0}{v}_2 + a_1t_1\overset{0}{v}_2\overset{1}{v}_{1,1} + a_2\overset{1}{v}_{2,1}\overset{0}{v}_{2,2} + a_2\overset{0}{v}_{2,2}\overset{1}{v}_{2,1} \\
& \qquad + c_1\overset{1}{v}_1\overset{0}{v}_{3,2} + c_1\overset{0}{v}_{3,2}\overset{1}{v}_1 - a_2b_2\overset{0}{v}_{2,1}\overset{1}{v}_3 - a_2b_2\overset{1}{v}_3\overset{0}{v}_{2,1}\Big] \\
& + \frac{1}{2}\Big[(c_1c_2 + a_2t_1t_2)\,\overset{0}{v}_1\overset{1}{v}_2 + (c_1c_2 + a_2t_1t_2)\,\overset{1}{v}_2\overset{0}{v}_1 - a_2b_2\overset{1}{v}_{2,1}\overset{0}{v}_3 - a_2b_2\overset{0}{v}_3\overset{1}{v}_{2,1} \\
& \qquad - (a_2b_2t_1 + a_1b_1t_1)\,\overset{0}{v}_2\overset{1}{v}_3 - (a_2b_2t_1 + a_1b_1t_1)\,\overset{1}{v}_3\overset{0}{v}_2\Big] \\
& + \frac{1}{2}\Big[-a_2t_1b_2\overset{1}{v}_2\overset{0}{v}_3 - a_2t_1b_2\overset{0}{v}_3\overset{1}{v}_2 + c_2\overset{1}{v}_{3,1}\overset{0}{v}_2 + c_2\overset{0}{v}_2\overset{1}{v}_{3,1}\Big] \\
& + \frac{1}{2}\Big[a_2t_1\overset{1}{v}_2\overset{0}{v}_{2,2} + a_2t_1\overset{0}{v}_{2,2}\overset{1}{v}_2\Big]
\end{aligned}
\tag{B.6}
$$

$$
\begin{aligned}
\overset{2}{\varepsilon}_{11} = {} & -a_1b_1\overset{1}{v}_{1,1} + a_1(b_1)^2\overset{1}{v}_3 \quad \text{(LIN5)} \\
& + \frac{1}{2}\Big[a_1(\overset{1}{v}_{1,1})^2 + a_2(\overset{1}{v}_{2,1})^2 + a_2(t_1)^2(\overset{1}{v}_2)^2 + (c_1)^2(\overset{1}{v}_1)^2 + a_1(b_1)^2(\overset{1}{v}_3)^2 + (\overset{1}{v}_{3,1})^2\Big] \\
& + \frac{1}{2}\Big[a_2t_1\overset{1}{v}_{2,1}\overset{1}{v}_2 + a_2t_1\overset{1}{v}_2\overset{1}{v}_{2,1} - a_1b_1\overset{1}{v}_{1,1}\overset{1}{v}_3 - a_1b_1\overset{1}{v}_3\overset{1}{v}_{1,1} + c_1\overset{1}{v}_{3,1}\overset{1}{v}_1 + c_1\overset{1}{v}_1\overset{1}{v}_{3,1}\Big]
\end{aligned}
\tag{B.7}
$$

$$
\begin{aligned}
\overset{2}{\varepsilon}_{22} = {} & -a_2b_2\overset{1}{v}_{2,2} - a_2b_2t_2\overset{1}{v}_1 + a_2(b_2)^2\overset{1}{v}_3 \quad \text{(LIN5)} \\
& + \frac{1}{2}\Big[a_1(\overset{1}{v}_{1,2})^2 + \big(a_1(t_1)^2 + (c_2)^2\big)\,(\overset{1}{v}_2)^2 + a_2(\overset{1}{v}_{2,2})^2 + a_2(t_2)^2(\overset{1}{v}_1)^2 \\
& \qquad + a_2(b_2)^2(\overset{1}{v}_3)^2 + (\overset{1}{v}_{3,2})^2\Big] \\
& + \frac{1}{2}\Big[a_1t_1\overset{1}{v}_{1,2}\overset{1}{v}_2 + a_1t_1\overset{1}{v}_2\overset{1}{v}_{1,2} + a_2t_2\overset{1}{v}_{2,2}\overset{1}{v}_1 + a_2t_2\overset{1}{v}_1\overset{1}{v}_{2,2}\Big] \\
& + \frac{1}{2}\Big[-a_2b_2\overset{1}{v}_{2,2}\overset{1}{v}_3 - a_2b_2\overset{1}{v}_3\overset{1}{v}_{2,2} - c_2\overset{1}{v}_{3,2}\overset{1}{v}_2 - c_2\overset{1}{v}_2\overset{1}{v}_{3,2}\Big] \\
& + \frac{1}{2}\Big[-a_2b_2t_2\overset{1}{v}_1\overset{1}{v}_3 - a_2b_2t_2\overset{1}{v}_3\overset{1}{v}_1\Big]
\end{aligned}
\tag{B.8}
$$

$$
\begin{aligned}
2\overset{2}{\varepsilon}_{12} = & -a_2 b_2 \overset{1}{v}_{2,1} - (a_2 b_2 t_1 + a_1 b_1 t_1)\, \overset{1}{v}_2 - a_1 b_1 \overset{1}{v}_{1,2} \quad \text{(LIN5)} \\
& + \frac{1}{2}\Big[a_1 \overset{1}{v}_{1,1} \overset{1}{v}_{1,2} + a_1 \overset{1}{v}_{1,2} \overset{1}{v}_{1,1} + a_2 \overset{1}{v}_{2,1} \overset{1}{v}_{2,2} + a_2 \overset{1}{v}_{2,2} \overset{1}{v}_{2,1} + (a_2 t_1 t_2 + c_1 c_2)\, \overset{1}{v}_2 \overset{1}{v}_1 \\
& \qquad + (a_2 t_1 t_2 + c_1 c_2)\, \overset{1}{v}_1 \overset{1}{v}_2 + \overset{1}{v}_{3,1} \overset{1}{v}_{3,2} + \overset{1}{v}_{3,2} \overset{1}{v}_{3,1} \Big] \\
& + \frac{1}{2}\Big[a_1 t_1 \overset{1}{v}_{1,1} \overset{1}{v}_2 + a_1 t_1 \overset{1}{v}_2 \overset{1}{v}_{1,1} + a_2 t_2 \overset{1}{v}_{2,1} \overset{1}{v}_1 + a_2 t_2 \overset{1}{v}_1 \overset{1}{v}_{2,1} - a_1 b_1 \overset{1}{v}_3 \overset{1}{v}_{1,2} - a_1 b_1 \overset{1}{v}_{1,2} \overset{1}{v}_3 \Big] \\
& + \frac{1}{2}\Big[a_2 t_1 \overset{1}{v}_2 \overset{1}{v}_{2,2} + a_2 t_1 \overset{1}{v}_{2,2} \overset{1}{v}_2 - a_2 b_2 \overset{1}{v}_{2,1} \overset{1}{v}_3 - a_2 b_2 \overset{1}{v}_3 \overset{1}{v}_{2,1} + c_1 \overset{1}{v}_1 \overset{1}{v}_{3,2} + c_1 \overset{1}{v}_{3,2} \overset{1}{v}_1 \Big] \\
& + \frac{1}{2}\Big[- (a_1 b_1 t_1 + a_2 b_2 t_1)\, \overset{1}{v}_3 \overset{1}{v}_2 - (a_1 b_1 t_1 + a_2 b_2 t_1)\, \overset{1}{v}_2 \overset{1}{v}_3 \Big] \\
& + \frac{1}{2}\Big[c_2 \overset{1}{v}_{3,1} \overset{1}{v}_2 + c_2 \overset{1}{v}_2 \overset{1}{v}_{3,1} \Big]
\end{aligned}
\tag{B.9}
$$

$$
\begin{aligned}
2\overset{0}{\varepsilon}_{23} = & \overset{1}{v}_2 + \overset{0}{v}_{3,2} + c_2 \overset{0}{v}_2 \quad \text{(LIN5)} \\
& + \frac{1}{2}\Big[a_1 \overset{0}{v}_{1,2} \overset{1}{v}_1 + a_1 \overset{1}{v}_1 \overset{0}{v}_{1,2} + a_2 \overset{0}{v}_{2,2} \overset{1}{v}_2 + a_2 \overset{1}{v}_2 \overset{0}{v}_{2,2} + \overset{0}{v}_{3,2} \overset{1}{v}_3 + \overset{1}{v}_3 \overset{0}{v}_{3,2} \Big] \\
& + \frac{1}{2}\Big[a_1 t_1 \overset{0}{v}_2 \overset{1}{v}_1 + a_1 t_1 \overset{1}{v}_1 \overset{0}{v}_2 + a_2 t_2 \overset{0}{v}_1 \overset{1}{v}_2 + a_2 t_2 \overset{1}{v}_2 \overset{0}{v}_1 \Big] \\
& + \frac{1}{2}\Big[-a_2 b_2 \overset{0}{v}_3 \overset{1}{v}_2 - a_2 b_2 \overset{1}{v}_2 \overset{0}{v}_3 + c_2 \overset{0}{v}_2 \overset{1}{v}_3 + c_2 \overset{1}{v}_3 \overset{0}{v}_2 \Big]
\end{aligned}
\tag{B.10}
$$

$$
\begin{aligned}
2\overset{0}{\varepsilon}_{13} = & \overset{1}{v}_1 + \overset{0}{v}_{3,1} + c_1 \overset{0}{v}_1 \quad \text{(LIN5)} \\
& + \frac{1}{2}\Big[a_1 \overset{0}{v}_{1,1} \overset{1}{v}_1 + a_1 \overset{1}{v}_1 \overset{0}{v}_{1,1} + a_2 \overset{0}{v}_{2,1} \overset{1}{v}_2 + a_2 \overset{1}{v}_2 \overset{0}{v}_{2,1} + \overset{0}{v}_{3,1} \overset{1}{v}_3 + \overset{1}{v}_3 \overset{0}{v}_{3,1} \Big] \\
& + \frac{1}{2}\Big[-a_1 b_1 \overset{0}{v}_3 \overset{1}{v}_1 - a_1 b_1 \overset{1}{v}_1 \overset{0}{v}_3 + a_2 t_1 \overset{0}{v}_2 \overset{1}{v}_2 + a_2 t_1 \overset{1}{v}_2 \overset{0}{v}_2 + c_1 \overset{0}{v}_1 \overset{1}{v}_3 + c_1 \overset{1}{v}_3 \overset{0}{v}_1 \Big]
\end{aligned}
\tag{B.11}
$$

$$
\begin{aligned}
2\overset{1}{\varepsilon}_{23} = & (c_2 - a_2 b_2)\overset{1}{v}_2 + \overset{1}{v}_{3,2} \quad \text{(LIN5)} \\
& + \frac{1}{2}\Big[a_1 \overset{1}{v}_{1,2} \overset{1}{v}_1 + a_1 \overset{1}{v}_1 \overset{1}{v}_{1,2} + a_2 \overset{1}{v}_{2,2} \overset{1}{v}_2 + a_2 \overset{1}{v}_2 \overset{1}{v}_{2,2} + \overset{1}{v}_{3,2} \overset{1}{v}_3 + \overset{1}{v}_3 \overset{1}{v}_{3,2} \Big] \\
& + \frac{1}{2}\Big[(a_1 t_1 + a_2 t_2)\overset{1}{v}_1 \overset{1}{v}_2 + (a_1 t_1 + a_2 t_2)\overset{1}{v}_2 \overset{1}{v}_1 \Big] \\
& + \frac{1}{2}\Big[(c_2 - a_2 b_2)\overset{1}{v}_2 \overset{1}{v}_3 + (c_2 - a_2 b_2)\overset{1}{v}_3 \overset{1}{v}_2 \Big]
\end{aligned}
\tag{B.12}
$$

$$
\begin{aligned}
2\overset{1}{\varepsilon}_{13} &= (c_1 - a_1 b_1)\overset{1}{v}_1 + \overset{1}{v}_{3,1} \quad \text{(LIN5)} \\
&+ \frac{1}{2}\left[a_1 \overset{1}{v}_{1,1}\overset{1}{v}_1 + a_1 \overset{1}{v}_1 \overset{1}{v}_{1,1} + a_2 \overset{1}{v}_{2,1}\overset{1}{v}_2 + a_2 \overset{1}{v}_2 \overset{1}{v}_{2,1} + \overset{1}{v}_{3,1}\overset{1}{v}_3 + \overset{1}{v}_3 \overset{1}{v}_{3,1}\right] \\
&+ \frac{1}{2}\left[2 a_2 t_1 (\overset{1}{v}_2)^2 + (c_1 - a_1 b_1)\overset{1}{v}_1 \overset{1}{v}_3 + (c_1 - a_1 b_1)\overset{1}{v}_3 \overset{1}{v}_1\right]
\end{aligned}
\tag{B.13}
$$

Appendix C
Normalization

C.1 Physical Components of the Strains

The relations between the original and normalized components of the Green strain tensor in the three defined curvilinear coordinate systems are shown in Table C.1, with the general coefficients in Table C.2.

Table C.1 Physical quantities of the Green strains

	Sphere	Cylinder	Plate	General
$\hat{\overset{n}{\varepsilon}}_{11}$	$\frac{R^2}{(R+\Theta^3)^2}\overset{n}{\varepsilon}_{11}$	$\overset{n}{\varepsilon}_{11}$		$s_1^2\overset{n}{\varepsilon}_{11}$
$\hat{\overset{n}{\varepsilon}}_{22}$	$\frac{1}{(R+\Theta^3)^2\sin^2\left(\frac{\Theta^1}{R}\right)}\overset{n}{\varepsilon}_{22}$	$\frac{1}{(R+\Theta^3)^2}\overset{n}{\varepsilon}_{22}$	$\overset{n}{\varepsilon}_{22}$	$s_2^2\overset{n}{\varepsilon}_{22}$
$\hat{\overset{n}{\varepsilon}}_{12}$	$\frac{R}{(R+\Theta^3)^2\sin\left(\frac{\Theta^1}{R}\right)}\overset{n}{\varepsilon}_{12}$	$\frac{1}{(R+\Theta^3)}\overset{n}{\varepsilon}_{12}$	$\overset{n}{\varepsilon}_{12}$	$s_1 s_2\overset{n}{\varepsilon}_{12}$
$\hat{\overset{n}{\varepsilon}}_{13}$	$\frac{R}{R+\Theta^3}\overset{n}{\varepsilon}_{13}$	$\overset{n}{\varepsilon}_{13}$		$s_1\overset{n}{\varepsilon}_{13}$
$\hat{\overset{n}{\varepsilon}}_{23}$	$\frac{1}{(R+\Theta^3)\sin\left(\frac{\Theta^1}{R}\right)}\overset{n}{\varepsilon}_{23}$	$\frac{1}{(R+\Theta^3)}\overset{n}{\varepsilon}_{23}$	$\overset{n}{\varepsilon}_{23}$	$s_2\overset{n}{\varepsilon}_{23}$

S.-Q. Zhang, *Nonlinear Analysis of Thin-Walled Smart Structures*, Springer Tracts in Mechanical Engineering, https://doi.org/10.1007/978-981-15-9857-9

Table C.2 Coefficients for the normalized strains

	Sphere	Cylinder	Plate	General
$s_1 = \|\mathbf{g}^1\|$	$\frac{R}{R+\Theta^3}$	1	1	$\sqrt{g^{11}}$
$s_2 = \|\mathbf{g}^2\|$	$\frac{1}{(R+\Theta^3)\sin\left(\frac{\Theta^1}{R}\right)}$	$\frac{1}{R+\Theta^3}$	1	$\sqrt{g^{22}}$

C.2 Physical Components of the Displacements

The relations between the original and normalized components of the displacement vector in the three defined curvilinear coordinate systems are shown in Table C.3, with the general coefficients in Table C.4.

Table C.3 Physical quantities of the displacements

	Sphere	Cylinder	Plate	General
$\overset{0}{v}_{1,1}$	$\overset{\hat{0}}{v}_{1,1}$			$\overset{\hat{0}}{v}_{1,1}$
$\overset{0}{v}_{1,2}$	$\overset{\hat{0}}{v}_{1,2}$			$\overset{\hat{0}}{v}_{1,2}$
$\overset{0}{v}_{2,1}$	$R\sin\left(\frac{\Theta^1}{R}\right)\overset{\hat{0}}{v}_{2,1}+$ $\cos\left(\frac{\Theta^1}{R}\right)\overset{\hat{0}}{v}_{2}$	$R\overset{\hat{0}}{v}_{2,1}$	$\overset{\hat{0}}{v}_{2,1}$	$k_1\overset{\hat{0}}{v}_{2,1}+k_2\overset{\hat{0}}{v}_{2}$
$\overset{0}{v}_{2,2}$	$R\sin\left(\frac{\Theta^1}{R}\right)\overset{\hat{0}}{v}_{2,2}$	$R\overset{\hat{0}}{v}_{2,2}$	$\overset{\hat{0}}{v}_{2,2}$	$k_1\overset{\hat{0}}{v}_{2,2}$
$\overset{0}{v}_{3,1}$	$\overset{\hat{0}}{v}_{3,1}$			$\overset{\hat{0}}{v}_{3,1}$
$\overset{0}{v}_{3,2}$	$\overset{\hat{0}}{v}_{3,2}$			$\overset{\hat{0}}{v}_{3,2}$
$\overset{1}{v}_{1,1}$	$\overset{\hat{1}}{v}_{1,1}$			$\overset{\hat{1}}{v}_{1,1}$
$\overset{1}{v}_{1,2}$	$\overset{\hat{1}}{v}_{1,2}$			$\overset{\hat{1}}{v}_{1,2}$
$\overset{1}{v}_{2,1}$	$R\sin\left(\frac{\Theta^1}{R}\right)\overset{\hat{1}}{v}_{2,1}+$ $\cos\left(\frac{\Theta^1}{R}\right)\overset{\hat{1}}{v}_{2}$	$R\overset{\hat{1}}{v}_{2,1}$	$\overset{\hat{1}}{v}_{2,1}$	$k_1\overset{\hat{1}}{v}_{2,1}+k_2\overset{\hat{1}}{v}_{2}$
$\overset{1}{v}_{2,2}$	$R\sin\left(\frac{\Theta^1}{R}\right)\overset{\hat{1}}{v}_{2,2}$	$R\overset{\hat{1}}{v}_{2,2}$	$\overset{\hat{1}}{v}_{2,2}$	$k_1\overset{\hat{1}}{v}_{2,2}$
$\overset{1}{v}_{3,1}$	$\overset{\hat{1}}{v}_{3,1}$			$\overset{\hat{1}}{v}_{3,1}$
$\overset{1}{v}_{3,2}$	$\overset{\hat{1}}{v}_{3,2}$			$\overset{\hat{1}}{v}_{3,2}$
$\overset{0}{v}_{1}$	$\overset{\hat{0}}{v}_{1}$			$\overset{\hat{0}}{v}_{1}$
$\overset{0}{v}_{2}$	$R\sin\left(\frac{\Theta^1}{R}\right)\overset{\hat{0}}{v}_{2}$	$R\overset{\hat{0}}{v}_{2}$	$\overset{\hat{0}}{v}_{2}$	$k_1\overset{\hat{0}}{v}_{2}$
$\overset{0}{v}_{3}$	$\overset{\hat{0}}{v}_{3}$			$\overset{\hat{0}}{v}_{3}$
$\overset{1}{v}_{1}$	$\overset{\hat{1}}{v}_{1}$			$\overset{\hat{1}}{v}_{1}$
$\overset{1}{v}_{2}$	$R\sin\left(\frac{\Theta^1}{R}\right)\overset{\hat{1}}{v}_{2}$	$R\overset{\hat{1}}{v}_{2}$	$\overset{\hat{1}}{v}_{2}$	$k_1\overset{\hat{1}}{v}_{2}$
$\overset{1}{v}_{3}$	$\overset{\hat{1}}{v}_{3}$			$\overset{\hat{1}}{v}_{3}$

Table C.4 Coefficients for the normalized displacements

	Sphere	Cylinder	Plate
k_1	$R\sin\left(\frac{\Theta^1}{R}\right)$	R	1
k_2	$\cos\left(\frac{\Theta^1}{R}\right)$	0	0